THE EARLY MATHEMATICAL MANUSCRIPTS OF LEIBNIZ

TRANSLATED FROM THE LATIN TEXTS
PUBLISHED BY CARL IMMANUEL GERHARDT
WITH CRITICAL AND HISTORICAL NOTES

BY

J. M. CHILD

MERCHANT BOOKS

PREFACE.

A STUDY of the early mathematical work of Leibniz seems to be of importance for at least two reasons. In the first place, Leibniz was certainly not alone among great men in presenting in his early work almost all the important mathematical ideas contained in his mature work. In the second place, the main ideas of his philosophy are to be attributed to his mathematical work, and not *vice versa*. The manuscripts of Leibniz, which have been preserved with such great care in the Royal Library at Hanover, show, perhaps more clearly than his published work, the great importance which Leibniz attached to suitable notation in mathematics and, it may be added, in logic generally. He was, perhaps, the earliest to realize fully and correctly the important influence of a calculus on discovery. The almost mechanical operations which we go through when we are using a calculus enable us to discover facts of mathematics or logic without any of that expenditure of the energy of thought which is so necessary when we are dealing with a department of knowledge that has not yet been reduced to the domain of operation of a calculus. There is a frivolous objection raised by philosophers of a superficial type, to the effect that such economy of thought is an attempt to substitute unthinking mechanism for living thought. This contention fails of its purpose through the simple fact that this economy is only used in certain circumstances. In no science do we try to make subject to a mechanical calculus any trains of reasoning except such that have not been the object of careful thought many times previously. Not only so, but this reasoning has been universally recognized as valid, and we do not wish to waste energy of thought in repeating it when so much remains to be discovered by means of this energy. Since the time of Leibniz, this truth has been recognized, explicitly or implicitly, by all the greatest mathematical analysts.

It is not difficult to connect with this great idea of the importance of a calculus in assisting deduction the many unfinished plans of Leibniz; for instance, his projects for an encyclopædia of all science, of a general science, of a calculus of logic, and so on. These projects, however, do not come within the field of this essay, which is a collection of various articles which appeared in *The Monist* from 1916 to 1918; our concern will be the various influences on Leibniz in his earliest original mathematical work. Merely biographical details do not seem to be relevant.

In writing the following pages, I have been greatly influenced and helped by the emphasis laid by Mr. Philip E. B. Jourdain upon the importance which Leibniz himself attached to the notion of a calculus in general, and his own operational calculus in particular; he it was who also suggested that I should undertake a critical translation of the early mathematical manuscripts of Leibniz; to him also I am greatly indebted for many points upon which I was unable to make up my mind on the evidence that I could get from the manuscripts alone. I have also to thank Mr. W. J. Greenstreet for looking through my articles before they were assembled for the purpose of this volume, and for making some valuable suggestions. My excuse for publishing these manuscripts, enlarged with so many and such long critical notes, must lie in the fact that I have made a careful study of the work of Barrow, and have recognized, perhaps at more than its true value, though I do not think so personally, its great genius and the influence it had on Leibniz. The opportunities it was capable of affording to Leibniz, the greater likeness that the work of Leibniz bears to that of Barrow than to that of Newton, have forced me to the conclusion that Leibniz was in no way indebted to Newton for anything, yet his statement in a letter to the Marquis d'Hospital, that he was under no obligation to Barrow *for his methods,* is absolutely correct.

J. M. CHILD.

DERBY, ENGLAND, September, 1919.

MERCHANT BOOKS

CONTENTS.

ISBN 1-60386-023-1

I.

INTRODUCTION.

APART from the intrinsic interest which the autograph writings, and more particularly the earlier efforts, of any of the prime movers in any branch of learning possess for the historical student, there is a special interest attached to the manuscripts and correspondence of Leibniz. They are invaluable as an aid to the study of the part that their author played in the invention and development of the infinitesimal calculus. More especially is this so in the case of Leibniz; for the matter, upon which this essay is founded, unearthed by Dr. C. I. Gerhardt in a mass of papers belonging to Leibniz that had been preserved in the Royal Library of Hanover, contained holographs previously unpublished.

The most important of these, for our purpose, were edited, with full notes and a commentary, by Gerhardt, in three separate volumes, under the respective titles:

1. *Historia et Origo Calculi Differentialis, a G. G. Leibnizio conscripta.* Hanover, 1846.

2. *Die Entdeckung der Differentialrechnung durch Leibniz.* Halle, 1848.

3. *Die Geschichte der höheren Analysis; erste Abtheilung, Die Entdeckung der höheren Analysis.* Halle, 1855.*

* For abbreviations used in this volume for these and other works, see the Bibliography given at the end.

The present time,[1] the two-hundredth anniversary of the death of Leibniz, would seem to be a most suitable one for publishing an English translation of these manuscripts.

For the present purpose, it will be convenient to group the manuscripts in two sections, of which the first will consist of Leibniz's own account of his work. Under the heading § 1, (p. 11), is given a fairly literal translation of a postscript from Leibniz to Jakob (i. e., James) Bernoulli, "which was written from Berlin in April 1703, and then cancelled and a postscript on a totally different subject substituted."[2] This is a communication to a more or less intimate friend. It is therefore naturally not such a considered composition as the second account that Leibniz gives of his work in the *Historia* mentioned above, of which a full translation is given below under the heading § 2. It is important to bear this point in mind when comparing the two accounts together, for any slight discrepancies that may be noticed are, feasibly at least, to be accounted for by the different circumstances of the compositions. The latter account bears the impress of being fairly fully revised and made ready for press, and the facts marshalled to make an impressive or, as some would have it, plausible whole; it was probably finished just before the death of Leibniz, and represents his answer to the *Commercium Epistolicum* of unsavory memory. The death of Leibniz in November 1716 was probably the cause which prevented its publication, or at least the chief reason.

It is not my intention to enter into a discussion about the *Commercium Epistolicum*; this has probably had the last word said upon it that it is possible to say with the help of the existing authentic material that is possessed by the present-day historians of mathematics. Further,

---

[1] This appeared in *The Monist* for October, 1916.

[2] G. 1848, p. 29; see also G. *math.*, III, pp. 71, 72, and Cantor, III, p. 40.

I hold quite other views as to the possible source of Leibniz's inspiration, if indeed he is not to be credited with perfectly independent discovery. I will therefore, as far as I may, refrain from allusion to the *Commercium Epistolicum,* except to second the plea of its perfectly disgraceful unfairness, as made by Leibniz.[3] I have suggested above that the *Historia* was intended by Leibniz as a statement of his side of the case, and as an answer to the attack made upon him. This account of his work, although written in the third person, "by a friend who knew all about the matter,"[4] is, on the authority of Gerhardt, undoubtedly by Leibniz himself. Without in any way impugning this authority, I cannot help thinking it would have been more satisfactory if I could have included herein photographic copies of parts of this manuscript; but this is impossible at the time of writing.

The reasons for the delay in the preparation of the *Historia* are stated in the manuscript itself; and later I shall have occasion to discuss these. In order that the remarks made may in all cases be perfectly intelligible, I must here give a very short account[5] of the history of the

---

[3] A fair-minded consideration, like everything emanating from the pen of De Morgan, is given of the matter in a recent edition of his *Essays on the Life and Work of Newton.* The tale is told with the charm characteristic of De Morgan, and the edition is rendered very valuable by the addition of notes, commentary, and a large number of references supplied by the editor, P. E. B. Jourdain (Open Court Publishing Co.). Special attention is directed to De Morgan's summary of the unfairness of the case in Note 3 at the foot of pages 27-28.

[4] See under 11 below: also cf. the original Latin as given in G. 1846, p. 4, *"per amicum conscium."*

[5] The account here given is substantially that given by Gerhardt in an article in Grunert's *Archiv der Mathematik und Physik,* 1856; pp. 125-132. This article is written in contradiction to the view taken by Weissenborn in his *Principien der höheren Analysis,* Halle, 1856. It is worthy of remark that the partisanship of Gerhardt makes him omit in this article all mention of the review which Leibniz wrote for the *Acta Eruditorum* on Newton's work, *De Quadratura Curvarum,* which really drew upon him the renewal of the attack, by Keill. The passage which was objected to by the English mathematicians as being tantamount to a charge of plagiarism, in addition to the insult implied, according to their thinking, in making Newton the fourth proportional to Cavalieri, Fabri and Leibniz, is however given by Gerhardt in his preface to the *Historia* (G. 1846, p. vii).

quarrel up to the time of the publication of the *Commercium Epistolicum* in 1712.

The matter was first started in the year 1699 by Fatio de Duillier, a Swiss mathematician who had been living in London since 1691; he was a correspondent of Huygens, and from letters that Fatio sent to Huygens[6] it would appear that the attack had been quietly in preparation for some time. Whether he had Newton's sanction or not cannot be ascertained, yet it seems certain from the correspondence that Newton had given Fatio information with regard to his writings. Fatio then concludes that Newton is the first discoverer and that Leibniz, as second discoverer, has borrowed from Newton. These accusations hurt Leibniz all the more, because he had deposited copies of his correspondence with Newton in the hands of Wallis for publication. As Fatio was a member of the Royal Society, Leibniz took it for granted that Fatio's attack was with the approval of that body; he asked therefore that the papers in the hands of Wallis should be published in justice to himself. He received a reply from Sloane, one of the secretaries of the Society, informing him that his assumption with regard to any such participation of the Society in the attack was groundless; and in consequence of this he took no further notice of the matter, and the whole thing lapsed into oblivion.

In the year 1708 the attack against Leibniz was renewed by Keill; and the charge that Leibniz had borrowed from Newton was most directly made. Leibniz had nobody in England who was in a position to substantiate his claims, for Wallis had died in 1703; so he appealed directly to the Royal Society. This body in consequence appointed a commission composed of members of the Society to consider the papers concerned in the matter. Their report

---

[6] Fatio's correspondence with Huygens is to be found in *Ch. Hugenii aliorumque seculi XVII virorum celebrium exercitationes mathematicae et philosophicae*, ed. Uylenbroeck, 1833.

appeared in the year 1712 under the title of *Commercium Epistolicum D. Johannis Collinsii et Aliorum de Analysi promota, jussu Societatis Regiae in lucem editum.*

Leibniz did not return to Hanover, from a tour of the towns of Italy on genealogical research work, until two years later; so that the date of the *Historia* is definitely established to have between 1714 and 1716, the date of his death. The dates allow us to account for the similarity between the two reports he gives of his work, in the postscript and the *Historia,* and also for any slight discrepancies between them.

Let us first, however, try to find a reason why the postscript was written, and having been written why it was cancelled. In the *Acta Eruditorum* (Leipsic) for January 1691, James Bernoulli said that Leibniz had got his fundamental ideas from Barrow;[7] but in a later number, that for June 1691, he admitted that Leibniz was far in advance of Barrow, though both views were alike in some ways.[8] One is inclined to wonder whether this admission was a result of Leibniz's reputed personality and charm; but as Leibniz seems to have been stationed at Wolfenbüttel and Bernoulli at Basel at this time a personal interview would seem improbable, and a more feasible suggestion would seem to be a reasoned remonstrance by letter from Leibniz. It is to be noticed that Bernoulli does not exactly retract his statement that Leibniz had Barrow to thank for the fundamental ideas, he only states that in spite of the similarities there are also dissimilarities in which Leibniz stands far above Barrow.[9] I am inclined to think he is simply comparing the method of Leibniz with the differential triangle method of Barrow, and that Bernoulli even has not noticed that Barrow has propositions that are the geometrical

---

[7] Bernoulli (Jakob), *Opera,* Vol. I, p. 431.

[8] *Ibid.,* p. 453.

[9] Cantor, III, p. 221.

equivalents of the differentiation of a product, quotient and powers of the *dependent* variables.

It seems to me that at this time Leibniz, though he does not forget his insinuation, has to lay all thoughts of combating it aside; for Gerhardt apparently found no other letters or other manuscripts referring to the matter prior to that of 1703. At a certain time later, judging from the first paragraph of the intended postscript, he would appear to have referred to the matter again, and to have called forth from the Bernoullis an excuse or a justification of the statements in the *Acta Eruditorum,* together with some expression of their surprise that he should have been upset over them. The reason may have been that it got to the ears of Leibniz that the opinion was not confined to the Bernoullis, for Leibniz says "....you, your brother, *or any one else.*"[10]

Thus much we may guess as to the occasion that prompted the writing of the postscript; now let us try to find the reason for its being cancelled. Fatio's attack seems to have been precipitated through pique at having been left out by Leibniz in a list of mathematicians alone capable of solving John Bernoulli's problem of the line of quickest descent.[11] "He published a memoir on the problem, in

---

[10] In the opening paragraph of the "postscript," page 11.

[11] The account which follows is taken from Williamson's article, "Infinitesimal Calculus," in the *Times* edition of the *Encyc. Brit.* The memoir referred to contains a passage, of which the following is a translation (G., 1846, p. v):

"Perhaps the distinguished Leibniz may wish to know how I came to be acquainted with the calculus that I employ. I found out for myself its general principles and most of the rules in the year 1687, about April and the months following, and thereafter in other years; and at the time I thought that nobody besides myself employed that kind of calculus. Nor would I have known any the less of it, if Leibniz had not yet been born. And so let him be lauded by other disciples, for it is certain that I cannot do so. This will be all the more obvious, if ever the letters which have passed between the distinguished Huygens and myself come to be published. However, driven thereto by the very evidence of things, I am bound to acknowledge that Newton was the first, and by many years the first, inventor of this calculus; from whom, whether Leibniz, the second inventor, borrowed anything, I prefer that the decision should lie, not with me, but with others who have had sight of the paper of Newton, and other additions to this same manuscript. Nor does

which he declared that he was obliged by the undeniable evidence of things to acknowledge Newton, not only as the first, but as by many years the first, inventor of the calculus; from whom, whether Leibniz, the second inventor, borrowed anything or not, he would rather *those who had seen Newton's letters and other manuscripts should judge than himself."* The attack in itself is cowardly, in that Fatio does not dare to make a direct assertion, only an insinuation that is far more damaging, since it suggests that to those who have seen the papers of Newton the matter could not be in the slightest doubt. Leibniz replied by an article in the *Acta Eruditorum,* for May 1700, in which he cited Newton's letters, as also the testimony which Newton rendered to him in the *Principia,*[12] as proof of his claim to an independent authorship of his method. A reply was sent by Duillier, which the editors of the *Acta Eruditorum* refused to publish. This would probably be in 1701; and I suggest that Leibniz had probably now come to the conclusion that it would be wiser to let the matter of Barrow drop and attend to the affair with Newton. When he, unwisely, started the controversy once more by a review (containing what was taken to be an implied sneering allusion to Newton) of the *Tractatus de Quadratura Curvarum,* published by the latter with his Optics in 1708,[13] and thus drew upon himself the attack

the silence of the more modest Newton, or the forward obtrusiveness of Leibniz...."

Truly another Roland in the field, and one in a vicious mood. What with other claimants to the method, such as Slusius, etc., at least as far as the differentiation of implicit functions of two variables is concerned, it would almost seem that the infinitesimal calculus was not an invention, but a *gradual* development of the fundamental principles of the ancient mathematicians.

[12] See De Morgan's *Newton,* p. 26 and pp. 148, 149, where the Scholium is translated. The original Latin of this Scholium to Lemma II of Book II of the *Principia,* the altered Scholium that appeared in the second and third editions, with a note remarking on the change, will be found on pp. 48, 49, in Book II of the "Jesuits' Edition" of Newton (Editio Nova, edited by J. M. F. Wright, Glasgow, 1822; the third and best edition of the work of Le Seur and Jacquier).

[13] *Phil. Trans.,* 1708; see also Cantor, III, p. 299.

by Keill, he gladly allowed the suggestion about Barrow to fade into oblivion, cast out by the more public, but I think the less true, charge of plagiarism from Newton. He also saw that he would have to prepare a careful answer if he made one at all, and second thoughts suggested that it would be as well if his postscript was made the matter for further consideration, correction, if necessary, and amplification, before it was sent off. It is to be noted that the review above mentioned is written anonymously in the third person, but it has been established that its author was Leibniz himself.[14]

There does not seem to be any occasion for further general remarks; particular points of criticism will be alluded to as the translation given below proceeds.

[14] For a discussion, see Rosenberger, *Isaac Newton und seine physikalischen Principien,* Leipsic, 1895.

II.

LETTER TO BERNOULLI.

§ 1.

Full translation of the intended postscript to the letter to James Bernoulli, dated April, 1703, from Berlin.

Perhaps[15] you will think it small-minded of me that I should be irritated with you, your brother, or any one else, if you should have perceived the opportunities for obligation to Barrow, which it was not necessary for me, his contemporary[16] in these discoveries, to have obtained from him.

When I arrived in Paris in the year 1672, I was self-taught as regards geometry,[17] and indeed had little knowledge of the subject, for which I had not the patience to read through the long series of proofs. As a youth I consulted the beginner's Algebra of a certain Lanzius,[18] and afterward that of Clavius;[19] that of

[15] The manner of the opening of this postscript would seem to indicate that something had been mentioned with regard to the matter of his irritation about imputed obligations to Barrow in the body of the letter; this cannot be ascertained, for Gerhardt does not quote the letter in connection.

[16] Leibniz can hardly with justice call Barrow his contemporary; Barrow anticipated him by half a dozen years at least. For Barrow had published his *Lectiones Geometricae* in 1670, while the very earliest date at which Leibniz could have obtained his results is the end of 1672; and there is reason to believe, as I have shown in my edition of the *Lectiones,* that Barrow was in possession of his method many years before publication, and had most probably communicated his secret to Newton in 1664.

[17] It is to be noted that the sole topic of this postscript is geometry, of which Leibniz candidly states that he knew practically nothing in 1672.

[18] Most probably the *Institutiones arithmeticae* of Johann Lantz, published at Munich in 1616; Cantor, III, p. 40.

[19] Possibly the *Geometria practica* of Christopher Clavius, better known as an editor of Euclid; he was the professor at Rome under whom Gregory St. Vincent studied. There are repeated references to Clavius in Cantor, II and III, Index, *q. v.*
It is worth remarking that neither Lanzius nor Clavius is mentioned in the *Historia.*

Descartes seemed to be more intricate.[20] Nevertheless, it seemed to me, I do not know by what rash confidence in my own ability, that I might become the equal of these if I so desired. I also had the audacity to look through even more profound works, such as the geometry of Cavalieri,[21] and the more pleasant elements of curves of Leotaud,[22] which I happened to come across in Nuremberg, and other things of the kind; from which it is clear that I was now ready to get along without help,[23] for I read them almost as one reads tales of romance.

Meanwhile I was fashioning for myself a kind of geometrical calculus by means of little squares and cubes to express undetermined numbers, being unaware that Descartes and Vieta had worked out the whole matter in a superior manner.[24] In this, I may almost call it, superb ignorance of mathematics, I was then studying history and law; for I had decided to devote myself to the latter. From mathematics I as it were only sipped those things that were the more pleasant, being especially fond of investigating and inventing machines, for it was at this time that my arithmetical

[20] It has been stated that, according to Descartes's own words, the intricacies of his *Géométrie* were intentional; it certainly has the character of a challenge to his contemporaries. There is no preparation, such as marks a book of the present day on coordinate geometry; Descartes starts straightway on the solution of a problem given up as insoluble by the ancients. No wonder that young Leibniz found some difficulty with his first attempt to read it.

[21] In 1635, Cavalieri published his *Geometria Indivisibilibus,* and thus laid the foundation stone of the integral calculus. It would seem that Roberval was really the first inventor, or at least an independent inventor of the method; but he lost credit for it because he did not publish it, preferring to keep it to himself for his own use. Other examples of this habit are common among the mathematicians of the time.

[22] The book referred to was published in 1654. It appeared as the second volume of a work whose first volume was a critique and refutation of the quadrature of the circle published by Gregory St. Vincent; this second volume was not the work of Leotaud, as the second part of the title showed: "necnon CURVILINEORUM CONTEMPLATIO, olim inita ab ARTUSIO DE LIONNE, Vapincensi Episc." It therefore appears to have been an edited reprint of the work of De Lionne, the bishop of Gap (ancient name, Vapincum). Since part of this treatise is devoted to the "lunules of Hippocrates" (see Cantor, I, pp. 192-194), it may have had some influence with Leibniz in giving him the first idea for his evaluation of π.

[23] Literally, "I was about to swim without corks."

[24] Leibniz here would appear to assert that he had considered some form of rectangular coordinate geometry, the association with the name of Descartes being fairly conclusive. Vieta's *In Artem Analyticam Isagoge* explained how algebra could be applied to the solution of geometrical problems (Rouse Ball); for further information see Cantor.

machine[25] was devised.   At this time also it happened that Huygens, who I fully believe saw more in me than there really was, with great courtesy brought me a copy recently published of his book on the pendulum.[26]   This was for me the beginning or occasion

[25] This seems to have been an improvement on the adding machine of Pascal, adapting it to multiplication, division and extraction of roots. Pascal's machine was produced in 1642, and Leibniz's in 1671.

[26] Huygens's *Horologium Oscillatorium* was published in 1673; we are thus provided with an exact date for the occurrence of the conversation that set Leibniz on to read Pascal and St. Vincent. This was after his first visit to London, from which he returned in March, "having utilized his stay in London to purchase a copy of Barrow's *Lectiones,* which Oldenburg had brought to his notice" (Zeuthen, *Geschichte der Mathematik im XVI. und XVII. Jahrhundert*; German edition by Mayer, p. 66). Leibniz himself mentions in a letter to Oldenburg, dated April 1673, that he has done so. Gerhardt (G. 1855, p. 48) states that he has seen, in the Royal Library of Hanover the copy of Barrow's *Lectiones Geometricae,* so that it must have been the combined edition of the Optics and the Geometry, published in 1670, that Leibniz bought.

Thus, before he is advised to study Pascal by Huygens, he has already in his possession a copy of Barrow. It is idle that any one should suppose that Leibniz bought this book on the recommendation of a friend in order merely to possess it; Leibniz bought books, or borrowed them, for the sole purpose of study. Unless we are to look upon this account of his reading as the result of lack of memory extending back for thirty years, there is only one conclusion to come to, barring of course the obviously brutal one that Leibniz lied; and this conclusion is that at the first reading the only thing that Leibniz could follow in Barrow was the part that he marked *Novi dudum* ("Knew this before"), and this was the appendix to Lecture XI, which dealt with the *Cyclometria* of Huygens, as Barrow calls the book entitled *De Circuli Magnitudine Inventa.* The absence of any more such remarks is almost proof positive that Leibniz knew none of the rest before. Hence he must have read the Barrow before he had filled those "hundreds of sheets" that he speaks of later, with geometrical theorems that he has discovered; for at the end of the postscript we are considering he states that "in Barrow, *when his Lectures appeared,* I found the greater part of my theorems anticipated." There is something very wrong somewhere; for this would appear to state that it was the second edition of Barrow, published in 1874, that Leibniz had bought; it is impossible, as the words of Leibniz stand, that they should refer to the 1670 edition, for it had been published before Leibniz arrived in Paris. It is however certain from Leibniz's letter to Oldenburg that it could not be the 1674 edition, for the date of the letter is 1673. In this letter Leibniz merely makes a remark on the optical portion; but it could not have been the separate edition of the Optics, published in 1669, for Gerhardt states that the copy he has seen contains the Geometry with notes in the margin.

To those who have ever waded through the combined edition of Barrow's Optics and Geometry, it may be that rather a startling suggestion will occur. It was sheer ill-luck that drove Leibniz, after studying the Optics (perhaps on the journey back from London, for we know that this was a habit of his), to get tired of the five preliminary geometrical lectures in all their dryness, and on reaching home, just to skim over the really important chapters, *missing all the important points,* and just the name of Huygens catching his eye. This is a new suggestion as far as I am aware; everybody seems to decide between one of two things, either that Leibniz never read the book until the date he himself gives, "*Anno Domini* 1675 as far as I remember," or else that he purposely lied. I will return to this point later; meanwhile see Cantor, III, pp. 161-163, and consult the references given in the footnotes to these pages; the pros and cons of the conflict between probability and Leibniz's word are there summarized.

of a more careful study of geometry. While we conversed, he perceived that I had not a correct notion of the center of gravity, and so he briefly described it to me; at the same time he added the information that Dettonville (i. e., Pascal) had worked such things out uncommonly well.[27] Now I, who always had the peculiarity that I was the most teachable of mortals, often cast aside innumerable meditations of mine that were not brought to maturity, when as it were they were swallowed up in the light shed upon them by a few words from some great man, immediately to grasp with avidity the teachings of a mathematician of the highest class; for I quickly saw how great was Huygens. In addition there was the stimulus of shame, in that I appeared to be ignorant with regard to such matters. So I sought a Dettonville from Buotius, a Gregory St. Vincent[28] from the Royal Library, and started to study geometry in earnest. Without delay I examined with delight the *"ductus"* of St. Vincent, and the *"ungulae"* begun by St. Vincent and developed by Pascal,[29] and those sums and sums of sums and solids formed

[27] Pascal's chief work on centers of gravity is in connection with the cycloid, and solids of revolution formed from it. His method was founded on the indivisibles of Cavalieri. His work was issued as a challenge to contemporaries under the assumed name of Amos Dettonville, and under the same name he published his own solutions, after solutions had been given by Huygens, Wallis, Wren and others.

[28] The method of *ductus plani in planum*, the leading or multiplication of a plane into a plane, employed by Gregory St. Vincent in the seventh book of his *Opus Geometricum* (1649) is practically on the same fundamental principle as the present method of finding the volume of a solid by integration. A simple explanation may be given by means of the figure of a quarter of a cone.

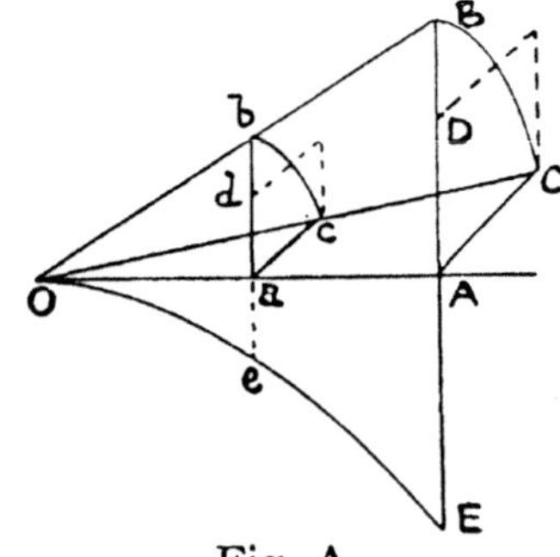

Fig. A.

Let AOBC be the quarter of a circular cone (Fig. A), of which OA is the axis, and ABC the base, so that all sections, such as *abc*, are parallel to ABC and perpendicular to the plane AOC. Let *ad* be the height of a rectangle equal in area to the quadrant *abc*, so that *ad* is the average height of the variable plane *abc*; then the volume of the figure is found by multiplying the height of the variable plane as it moves from O to the position ABC by the corresponding breadth of the plane OAC, i. e., by *bc*, and adding the results.

As we shall see later, Leibniz does not fully appreciate the real meaning of the method; on the other hand Wallis uses the method with good effect in his *Arithmetica Infinitorum*, and states that he has come to it independently. In the above case he would have stated that the product in each case was proportional to the square on *ac*, drawn an ordinate *ae* at right angles to O*a*, so that *ae* represented the product, and so formed the parabola OeEAaO, of which the area is known to him. This area is proportional to the volume of the cone.

[29] *Ungulae* denote hoof-shaped solids, such as the frusta of cylinders or cones cut off by planes that are not parallel to one another.

and resolved in various ways; for they afforded me more pleasure than trouble.

I was working upon these when I happened to come across a proof of Dettonville's that was of a supremely easy nature, by which he proved the mensuration of the sphere as given by Archi-

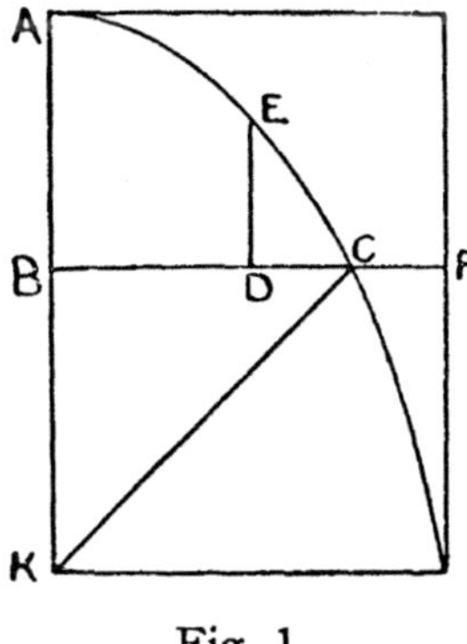

Fig. 1.

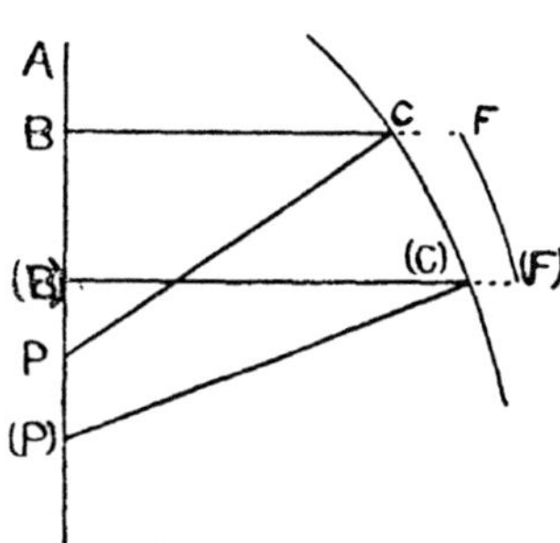

Fig. 2.

medes,[30] and showed from the similarity of the triangles EDC and CBK that CK into DE = BC into EC; and hence, by taking BF = CK,

[30] Figure 1 (see above) is of extreme interest. First of all it is not Barrow's "differential triangle," which is that of Fig. B below; this of course is only what those who believe Leibniz's statement that he received no help from Barrow, would expect. By the way, the figure given by Cantor as Barrow's is not quite accurate. (Cantor, III, p. 135.)

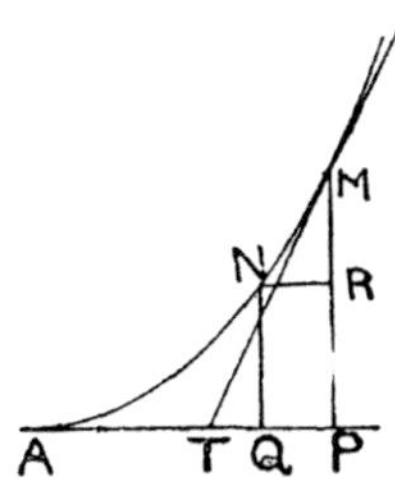

Fig. B (BARROW).

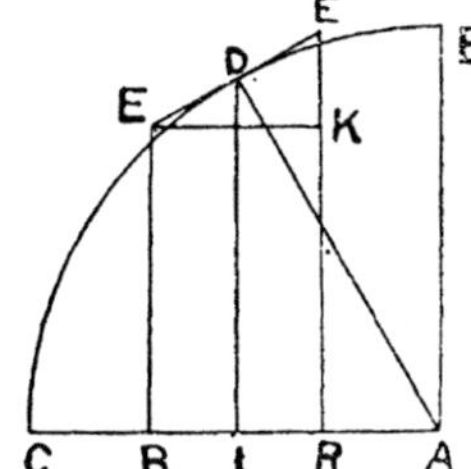

Fig. C (PASCAL).

But neither is it the figure of Pascal, which is that of Fig. C. Of course, I am assuming that Gerhardt has given a correct copy of the figure given by Leibniz in his manuscript; although that which I have given of it, a faithful copy of Gerhardt's, shows that his curve was not a circle. I also assume that Cantor is correct in the figure that he gives from Pascal; although Cantor says that the figure occurs in a tract on the sines of a quadrant, and not, as Leibniz states, in a problem on *the measurement of the sphere*. Indeed it seems to me that the figure is more likely to be connected with the area of the zone of a sphere and the proof that this is equal to the corresponding belt on the circumscribing cylinder than anything else. I am bound to assume these things, for I have not had the opportunity of seeing either of the figures in the original for myself. It is strange, in this connection, that Gerhardt in one place (G. 1848, p. 15) gives 1674 as the date of the publication of Barrow, and in another

that the rectangle AF is equal to the moment[31] of the curve AEC about the axis AB. [Fig. 1.]

place (G. 1855, p. 45) seven years later, he makes it 1672, and neither of them is correct as the date of the copy that Leibniz could possibly have purchased, namely 1670. This is culpable negligence in the case of a date upon which an argument has to be founded, for one can hardly suspect Gerhardt of deliberate intent to confuse. Nevertheless, like De Morgan, I should have felt more happy if I could have given facsimiles of Barrow's book, and Leibniz's manuscript and figure.

Lastly, there is in Barrow (what neither Gerhardt, Cantor, nor any one else, with the possible exception of Weissenborn, seems to have noticed) chapter and verse for Leibniz's "characteristic triangle." Fig. D is the diagram that Barrow gives to illustrate the first theorem of Lecture XI. This is of course, as is usual with Barrow, a complicated diagram drawn to do duty for a whole set of allied theorems.

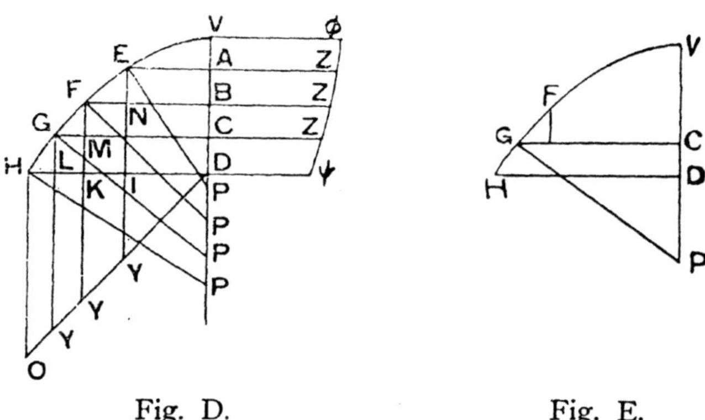

Fig. D.   Fig. E.

In the proof of the first of these theorems occur these words:
"Then the triangle HLG is similar to the triangle PDH (for, on account of the infinite section, the small arc HG can be considered as a straight line).
Hence, HL : LG = PD : DH, or HL . DH = LG . PD,
i. e., HL . HO = DC . Dψ.

By similar reasoning, it may be shown that, since the triangle GMF is similar to the triangle PCG,...."

If now the lines in italics are compared with that part of the figure to which they refer, which has been abstracted in Fig. E, the likeness to Leibniz's figure wants some explaining away, if we consider that Leibniz had the opportunity for seeing this diagram. Such evidence as that would be enough to hang a man, even in an English criminal court. (Further, see Note 46.)

To sum up, I am convinced that Leibniz was indebted to both of Barrow's diagrams, and also to that of Pascal (for I will call attention to the fact that he uses all three, as I come to them) and I think that after the lapse of thirty years he really could not tell from whom he got his figure. In such a case it would be only natural, if he knew that it was from one of two sources and he was accused of plagiarizing from the one, that he should assert that it was from the other. Hence, by repetition, he would come to believe it. But even this does not explain his letter to d'Hospital, where he says that he has not obtained any assistance from his methods; unless again we remember that this letter is dated 1694, twenty years after the event.

[31] Great importance, in my opinion hardly merited, is attached to the use by Leibniz of the phrase *momento ex axe* in this place, and in his manuscripts

The novelty of the reasoning struck me forcibly, for I had not noticed it in the works of Cavalieri.[32] But nothing astonished me so much as the fact that Pascal seemed to have had his eyes obscured by some evil fate; for I saw at a glance that the theorem was a most general one for any kind of curve whatever. Thus, let the perpendiculars not all meet in a point, but let each perpendicular from the

under the heading *Analysis Tetragonistica ex Centrobarycis,* dated October, 1675.

The Latin word *momentum,* a contraction of *movimentum,* has a primary meaning of movement or alteration, and a secondary meaning of a cause producing such movement. The present use of the term to denote the tendency of a force to produce rotation is an example of the use of the word to denote an effect; from the second idea, we have first of all its interpretation as something just sufficient to cause the alteration in the swing of a balance (where the primary idea still obtains), hence something very small, and especially a very small element of time.

Thus we see that Leibniz uses the term in its primary sense, for he employs it in connection with a method *ex Centrobarycis,* and in its mechanical sense, and it is thus fairly justifiable to assume that he got the term from Huygens; in just this sense we now speak of the moment of inertia.

Newton's use of the term is given in Lemma II of Book II of the *Principia,* in the following way.

"I shall here consider such quantities as undetermined or variable, as it were increasing or decreasing by a continual motion or flow (*fluxus*); and their instantaneous (*momentanea*) increments or decrements I shall denote (*intelligo* = understand) by the name "moments"; so that increments stand for moments that are added or positive (*affirmativis*), and decrements for those that are subtracted or negative."

This has nothing whatever to do with what Leibniz means by a moment, and it seem ridiculous to bring forward the use of this word as evidence that Leibniz had seen Newton's work, or even heard of it through Tschirnhaus, before the year 1675.

The fact that in another place, where I will refer to it again, he uses the phrase "instantaneous increment" is quite another matter.

The use of the word moment in this mechanical sense is here perfectly natural. See Cantor, III, p. 165; also Cantor, II, p. 569, where the *idea* is referred back at least to Benedetti (1530-1590); but the idea is fundamental in the theorems due to Pappus concerning the connection between the path of the center of gravity of an area and the surfaces and volumes of rings generated by the area, of which the proofs were given by Cavalieri. When, however, and by whom, the *word* moment was itself first used in this connection, I have been unable to find the slightest trace. (See p. 195.)

[32] With due regard to the statement that Leibniz "had looked through Cavalieri" before he went to Paris, it is not remarkable that he did not notice very much at all in Cavalieri. Cavalieri's *Geometria Indivisibilibus* is not a book to be "looked through." It is a work for weeks of study. I cannot say whether the idea involved in Leibniz's characteristic triangle is used by Cavalieri as such; but I do not see how else he could have given proofs (as stated by Williamson in his article on "Infinitesimal Calculus" in the *Times* edition of the *Encyc. Brit.*) of Pappus's theorem for the area of a ring; and I should think that it is morally certain that Cavalieri is the source from which Wallis obtained his ideas for the rectification of the arc of the spiral. I had occasion to refer to a copy in the Cambridge University Library, and what I saw of it in the short time at my disposal determined me to make a translation of it, with a commentary, *as soon as I had enough time at my disposal.* "As one reads tales of romance"!

curve be transferred to the position of an ordinate to the axis, as PC or (P)(C) to the position BF or (B)(F); then it is clear that the zone FB(B)(F)F will be equal to the moment of the curve C(C) about the axis.[33] [Fig. 2.]

Straightway I went to Huygens, whom I had not seen again in the meantime. I told him that I had followed out his instructions and that I was now able to do something that Pascal had failed to do. Then I showed him the general theorem for moments of curves. He was struck with wonder and said, "Now, that is the very theorem upon which depend my constructions for finding the area[34] of the surfaces of parabolic, elliptic and hyperbolic conoids; and how these were discovered, neither Roberval nor Bullialdus[35] were ever able to understand." Thus praising my progress, he asked me whether I could not now find the properties of such curves as F(F). When I told him that I had made no investigation in this direction he told me to read the works of Descartes and Slusius,[36] who showed how to form equations for loci; for he said that this idea was a most useful one. Thereupon I examined the Geometry of Descartes and made a close study of Slusius, thus entering the

[33] The moment is proportional to the area of the surface formed by the rotation of the curve C(C) about AP. Barrow does not at first use the method to find the areas of surfaces of revolution; he prefers to straighten out the curve C(C), and erect the ordinates BC, (B)(C) perpendicular to the curve thus straightened; i. e., he works with the product BC.C(C) as it stands. But, after giving the determination of the surface of a right circular cone as an example of the method, and as a means of combating the objections of Tacquet to the method of indivisibles, he goes on to say: "Evidently in the same manner we can investigate most easily the surfaces of spheres and portions of spheres (nay, provided all necessary things are given or known, any other surfaces that are produced in this way). But I propose to keep, to a great extent, to more general methods" (end of Lecture II). Thus we find that Barrow does not give any further examples of the determination of the areas of surfaces of revolution until Lecture XII. And why? Because he is not writing a work on mensuration, but a calculus. The reference to the method of indivisibles however shows that in Barrow's opinion, if Cavalieri had not used his method for the determination of the area of the surface of a sphere, then he ought to have done so.

[34] It is difficult to see also how Huygens could have performed his constructions unless he had used the method that Leibniz claims to have discovered.

[35] It is strange that Roberval, as an independent discoverer of the method of indivisibles, did not perceive the method of the constructions of Huygens. Bullialdus is Ismael Bouilleau (Martin's *Biog. Philos.*), or Boulliau (Poggendorff), author of works on conics, arithmetic of infinites, astronomy, etc. Cf. Seth Ward: *In Ismaelis Bullialdi astron. philos. fundamento inquisitiones.* Oxford, 1653.

[36] This conversation probably took place late in 1673; see a note on the alteration of the date of a manuscript dated November 11, 1673, where the 3 was originally a 5 (see p. 93).

house of geometry truly as it were by the back door. Urged on by the success I met with, and by the great number of results that I obtained, I filled some hundreds of sheets with them in that year. These I divided into two classes of assignables and inassignables. Among assignables I placed everything I obtained by the methods previously used by Cavalieri, Guldinus, Toricelli, Gregory St. Vincent and Pascal, such as sums, sums of sums, transpositions, "ductus," cylinders truncated by a plane, and lastly by the method of the center of gravity; and among inassignables I placed all that I obtained by the use of the triangle which I at that time called "the characteristic triangle,"[37] and things of the same class, of which Huygens and Wallis seemed to me to have been the originators.

A little later there fell into my hands the Universal Geometry of James Gregory of Scotland,[38] in which I saw the same idea ex-

The method of Slusius (de Sluze, or Sluse) is as follows:

Suppose that the equation of the given curve is

$$x^3 - 2x^2y + bx^2 - b^2x + by^2 - y^3 = 0.$$

Slusius takes all the terms containing y, multiplies each by the corresponding index of y; then similarly takes all the terms containing x, multiplies each by the corresponding index of x, and divides each term of the result by x; the quotient of the former by the last expression gives the value of the subtangent. This is practically the content of Newton's method of *analysis per aequationes*, and Slusius sent an account of it to the Royal Society in January, 1673. It was printed in the *Phil. Trans.*, as No. 90. This is given by Gerhardt (G. 1848, p. 15) as an example of the method of Slusius. It is rather peculiar that Gerhardt does not mention that this is the example given by Newton in the oft-quoted letter of December 10, 1672, and represents what Newton "guesses the method to be." As it stands in G. 1848, it would appear to be a quotation from the work of Slusius himself. There is evidence that Leibniz had seen the explanation given in the *Phil. Trans.*, or had been in communication with Slusius; this will be referred to later, but it may be said here that this fact makes Leibniz somewhat independent of any necessity of having seen Newton's letter.

[37] Some point is made of the question why, if Leibniz had seen the "differential triangle" of Barrow, he should have called it by a different name. If there were any point in it at all, it would go to prove that Barrow's calculus was published by Barrow as a *differential* calculus. But there is no point, for Barrow never uses the term! It is a product of later growth, by whom first applied I know not. Leibniz, thus free to follow his logical plan of denominating everything, uses a term borrowed from his other work. He thus defines a character or characteristic. "Characteristics are certain things by means of which the mutual relations of other things can be expressed, the latter being dealt with more easily than are the former." See Cantor, III, p. 33f.

[38] Gregory's *Geometriae Pars Universalis* was published at Padua in 1668. Leibniz had either this book, or the Barrow in which one of Gregory's theorems is quoted, close at hand in his work. For he gives it as an example of the power of his calculus, referring to a diagram which is not drawn. This diagram I was unable to draw from the meager description of it given by Leibniz, until *I looked up Barrow's figure*, in default of being able to obtain a copy of Gregory's work; thereupon the figure was drawn immediately.

ploited (although obscured by the proofs, which he gave according to the manner of the ancients), and as in Barrow, when his Lectures appeared, in which latter I found the greater part of my theorems anticipated.[39]

However I did not mind this very much, since I saw that these things were perfectly easy to the veriest beginner who had been trained to use them,[40] and because I perceived that there remained much higher matters, which however required a new kind of calculus. Thus I did not think that my Arithmetical Quadrature, although it was received by the French and English with great commendation, was worth being published, as I was loath to waste time over such trifles while the whole ocean was open to me. How matters then proceeded you already know, and as my letters, which the English themselves have published, prove.[41]

[39] Here indeed it must be admitted that Leibniz is —suffering from a lapse of memory. As has been said before, Barrow's lectures appeared in 1670 and were in the possession of Leibniz before ever he dreamed of his theorems. But what can one expect when admittedly this account (from which the *Historia* was in all probability written up) is purely from memory, aided by the few manuscripts that he had kept. Gerhardt does not say that he has found, nor does he publish, any manuscripts that could possibly give the order in which the text-books that Leibniz procured were read. Which of us, at the age of 57, could say in what order we had read books at the age of 27; or, if by then we had worked out a theory, could with accuracy describe the steps by which we climbed, or from a mass of muddle and inaccuracies, say to whom we were indebted for the first elementary ideas that we had improved beyond all recognition? I doubt whether any of us would recognize our own work under such circumstances.

[40] Again Leibniz makes a bad mistake in affecting to despise the work of his rivals—for that is what the words, "these things were perfectly easy to the veriest beginner who had been trained to use them," makes us believe. It is also bad taste, for, besides Barrow, Huygens also remained true to the method of geometry till his death. The sentence which follows savors of conceit; as a matter of fact it was left to others, such as the Bernoullis, to make the best use of the method of Leibniz. The great thing we have to thank Leibniz for is the notation; it is a mistake to call this the invention of a notation for the infinitesimal calculus. As we shall see, Leibniz invented this notation for finite differences, and only applied it to the case in which the differences were infinitely small. Barrow's method, of a and e, also survives to the present day, under the disguise of h and k, in the method by which the elements of the calculus are taught in nine cases out of ten. For higher differential coefficients the suffix notation is preferable, and later on the operator D is the method *par excellence*.

[41] Here Leibniz seems to be unable to keep from harking back to the charge made by Fatio, suggesting that by the publication of his letters by Wallis this charge has been proved to be absolutely groundless.

NOTE.

As I have pointed out in the Introduction, it is important, when comparing the foregoing "postscript" with the more detailed "Historia" which follows, the different circumstances of the compositions. Otherwise there is a danger that certain slight discrepancies between the two accounts may assume an importance that is not justified.

These discrepancies, however, have a certain amount of importance; especially in their relation to the indebtedness of Leibniz to Barrow rather than to Newton. The different diagrams, given by Leibniz in connection with his several explanations of the manner in which he obtained his "moment theorem," afford perhaps the greatest food for thought; and this, more especially perhaps in relation to the indebtedness of Leibniz to Pascal, which, in opposition to Gerhardt, I have tried to show was hardly worth mentioning. This point is discussed in notes on pp. 15-18, and further in Chap. VII. A second point is that mention is made of Barrow's *Lectiones* in the "postscript"; whereas the name of Barrow is omitted in the "Historia" from the list of those noteworthy men who dealt with indivisibles, given on p. 24. This is connected with the date of purchase by Leibniz of his copy of Barrow's book, and the incorrect date of publication given by Gerhardt.

Lastly, in the postscript there is only a passing mention made of Leibniz's Arithmetical Quadrature; whereas in the "Historia" (p. 42) it is given in great detail, showing the importance that Leibniz assigned to his method of transmutation. In a note on p. 172, I show that there was no necessity for Leibniz to have seen Newton's work on Series; for a straightforward application of Mercator's method of summation to a result given by Barrow yields the arc in terms of the tangent in the form usually known as Gregory's Series. We now proceed to consider the "Historia."

III.

"HISTORIA ET ORIGO CALCULI DIFFEREN-
TIALIS."

§ 2.

HISTORY AND ORIGIN OF THE DIFFERENTIAL CALCULUS.

It is an extremely useful thing to have knowledge of the true origins of memorable discoveries, especially those that have been found not by accident but by dint of meditation. It is not so much that thereby history may attribute to each man his own discoveries and that others should be encouraged to earn like commendation, as that the art of making discoveries should be extended by considering noteworthy examples of it.

Among the most renowned discoveries of the times must be considered that of a new kind of mathematical analysis, known by the name of the differential calculus; and of this, even if the essentials are at the present time considered to be sufficiently demonstrated, nevertheless the origin and the method of the discovery are not yet known to the world at large. Its author invented it nearly forty years ago, and nine years later (nearly thirty years ago) published it in a concise form; and from that time it has not only been frequently made known in memoirs,[42] but also has been a method of general employment; while many splendid discoveries have been made by its assistance, such as have been included in the *Acta Eruditorum*, Leipsic, and also such as have been published in the memoirs of the Royal Academy of Sciences; so that it would seem that a new aspect has been given to mathematical knowledge arising out of its discovery.

Now there never existed any uncertainty as to the name of the true inventor, until recently, in 1712, certain upstarts, either

---

[42] It is possible that this may mean "has received high commendation"; for *elogiis* may be the equivalent of eulogy, in which case *celebratus est* must be translated as "has been renowned."

in ignorance of the literature of the times gone by, or through envy, or with some slight hope of gaining notoriety by the discussion, or lastly from obsequious flattery, have set up a rival to him; and by their praise of this rival, the author has suffered no small disparagement in the matter, for the former has been credited with having known far more than is to be found in the subject under discussion. Moreover, in this they acted with considerable shrewdness, in that they put off starting the dispute until those who knew the circumstances, Huygens, Wallis, Tschirnhaus, and others, on whose testimony they could have been refuted, were all dead.[43] Indeed this is one good reason why contemporary prescripts should be introduced as a matter of law; for without any fault or deceit on the part of the responsible party, attacks may be deferred until the evidence with which he might be able to safeguard himself against his opponent had ceased to exist. Moreover, they have changed the whole point of the issue, for in their screed, in which under the title of *Commercium Epistolicum D. Johannis Collinsii* (1712) they have set forth their opinion in such a manner as to give a dubious credit to Leibniz, they have said very little about the calculus; instead, every other page is made up of what they call infinite series. Such things were first given as discoveries by Nicolaus Mercator[44] of Holstein, who obtained them by the process

[43] This is untrue. As has been said, the attack was first made publicly in 1699; at this time, although Huygens had indeed been dead for four years, Tschirnhaus was still alive, and Wallis was appealed to by Leibniz. It is strange that Leibniz did not also appeal to Tschirnhaus, through whom it is suggested by Weissenborn that Leibniz may have had information of Newton's discoveries. Perhaps this is the reason why he did not do so, since Tschirnhaus might not have turned out to be a suitable witness for the defense. Leibniz must have had this attack by Fatio in his mind, for he could hardly have referred to Keill as a *novus homo,* while we know that he did not think much of Fatio as a mathematician. To say that there never existed any uncertainty as to the name of the true inventor until 1712 is therefore sheer nonsense; for if by that he means to dismiss with contempt the attack of Fatio, whom can he mean by the phrase *novus homo?* The sneering allusion to "the hope of gaining notoriety by the discussion" can hardly allude to any one but Fatio. Finally if Fatio is dismissed as contemptible, the second attack by Keill was made in 1708. If it was early in the year, Tschirnhaus was even then alive, though Wallis was dead.

[44] Gerhardt says in a note (G. 1846, p. 22) that his real name was probably Kramer; for what reason I am unable to gather. Cantor says distinctly that his name was Kaufmann, and this is the usually accepted name of the man who was one of the first members of the Royal Society and contributed to its *Transactions.* It seems to me that Gerhardt is guessing; the German word *Kramer* means a small shopkeeper, while *Kaufmann* means a merchant. To Mercator is due the logarithmic series obtained by dividing unity by $(1 + x)$ and integrating the resulting series term by term; the connection with the logarithm of $(1 + x)$ is through the area of the rectangular hyperbola $y(1 + x) = 0$. See Reiff, *Geschichte der unendlichen Reihen.*

of division, and Newton gave the more general form by extraction
of roots.[45]   This is certainly a useful discovery, for by it arithmetical approximations are reduced to an analytical reckoning; but
it has nothing at all to do with the differential calculus. Moreover,
even in this they make use of fallacious reasoning; for whenever
this rival works out a quadrature by the addition of the parts by
which a figure is gradually increased,[46] at once they hail it as the
use of the differential calculus (as for instance on page 15 of the
Commercium).   By the selfsame argument, Kepler (in his *Stereometria Doliorum*),[47] Cavalieri, Fermat, Huygens, and Wallis used

[45] Newton obtained the general form of the binomial expansion after the
method of Wallis, i. e., by interpolation. See Reiff.

[46] We now see what was Leibniz's point; the differential calculus was
not the employment of an infinitesimal and a summation of such quantities;
it was the use of the idea of these infinitesimals being differences, and the
employment of the notation invented by himself, the rules that governed the
notation, and the fact that differentiation was the inverse of a summation;
and perhaps the greatest point of all was that the work had not to be referred
to a diagram. This is on an inestimably higher plane than the mere differentiation of an algebraic expression whose terms are simple powers and roots
of the independent variable.

[47] Why is Barrow omitted from this list? As I have suggested in the
case of Barrow's omission of all mention of Fermat, was Leibniz afraid to
awake afresh the sleeping suggestion as to his indebtedness to Barrow? I
have suggested that Leibniz read his Barrow on his journey back from London,
and perhaps, tiring at having read the Optics first and then the preliminary
five lectures, just glanced at the remainder and missed the main important
theorems. I also make another suggestion, namely, that perhaps, or probably,
in his then ignorance of geometry he did not understand Barrow. If this is
the case it would have been gall and wormwood for Leibniz to have ever
owned to it. Then let us suppose that in 1674 with a fairly competent knowledge of higher geometry he reads Barrow again, skipping the Optics of which
he had already formed a good opinion, and the wearisome preliminary lectures
of which he had already seen more than enough. He notes the theorems as
those he has himself already obtained, and the few that are strange to him
he translates into his own symbolism. I suggest that this is a feasible supposition, which would account for the marks that Gerhardt states are made in
the margin. It would account for the words "in which latter I found the
greater part of my theorems anticipated" (this occasion in future times rank-
ing as the first time that he had really read Barrow, and lapse of memory at
the end of thirty years making him forget the date of purchase, possibly confusing his two journeys to London); it would account for his using Barrow's
differential triangle instead of his own "characteristic triangle." As Barrow
tells his readers in his preface that "what these lectures bring forth, or to
what they may lead you may easily learn from *the beginnings of each*," let us
suppose that Leibniz took his advice. What do we find? The first four theorems of Lecture VIII give the geometrical equivalent of the differentiation
of a power of *a dependent* variable; the first five of Lecture IX lead to a
proof that, expressed in the differential notation,

$$(ds/dx)^2 = 1 + (dy/dx)^2;$$

the appendix to this lecture contains the differential triangle, and five examples on the *a* and *e* method, fully worked out; the first theorem in Lecture XI
has a diagram such that, when that part of it is dissected out (and Barrow's

the differential calculus; and indeed, of those who dealt with "indivisibles" or the "infinitely small," who did not use it? But Huygens, who as a matter of fact had some knowledge of the method of fluxions as far as they are known and used, had the fairness to acknowledge that a new light was shed upon geometry by this calculus, and that knowledge of things beyond the province of that science was wonderfully advanced by its use.

Now it certainly never entered the mind of any one else before Leibniz to institute the notation peculiar to the new calculus by which the imagination is freed from a perpetual reference to dia-

diagrams want this in most cases) which applies to a particular paragraph in the proof of the theorem, this portion of the figure is a *mirror image of the figure* drawn by Leibniz when describing the characteristic triangle (turn back to note 30). I shall have occasion to refer to this diagram again. The appendix to this lecture opens with the reference to the work of Huygens; and the second theorem of Lecture XII is the strangest coincidence of all. This theorem in Barrow's words is:

"Hence, if the curve AMB is rotated about the axis AD, the ratio of the surface produced to the space ADLK is that of the circumference of a circle to its diameter; whence, if the space ADLK is known, the said surface is known."

The diagram given by Barrow is as usual very complicated, serving for a group of nine propositions. Fig. F is that part of the figure which refers to the theorem given above, dissected out from Barrow's figure. Now remember that Leibniz always as far as possible kept his axis clear on the left-hand side of his diagram, while Barrow put his datum figure on the left of his axis, and his constructed figures on the right; then you have Leibniz's diagram and the proof is by the similarity of the triangles MNR, PMF, where

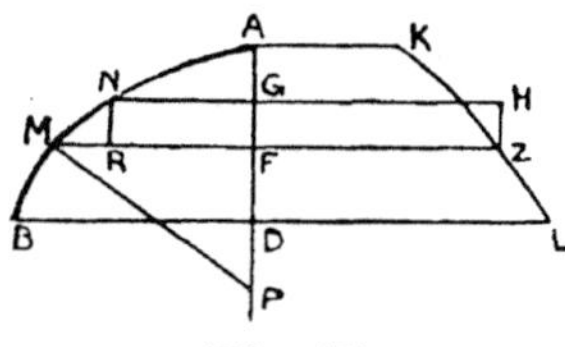

Fig. F.

FZ = PM; and the theorem itself is only another way of enunciating the theorem that Leibniz states he generalized from Pascal's particular case! Lastly, the next theorem starts with the words: "Hence the surfaces of the sphere, both the spheroids and the conoids receive measurement." What a coincidence!

As this note is getting rather long, I have given the full proof of the first two theorems of Barrow's Lecture XII as a supplement, at the end of this section.

The sixth theorem of this lecture is the theorem of Gregory which Leibniz also gives later; I will speak of this when I come to it. As also, when we discuss Leibniz's proof of the rules for a product, etc., I will point out where they are to be found in Barrow ready to his hand.

Yet if all this were so, he could still say with perfect truth that, in the matter of the invention of the differential calculus (as he conceived the matter to consist, that is, the differential and integral notations and the method of analysis), he derived no assistance from Barrow. In fact, once he had absorbed his fundamental ideas, Barrow would be less of a help than a hindrance.

grams, as was made by Vieta and Descartes in their ordinary or Apollonian geometry; moreover, the more advanced parts pertaining to Archimedean geometry, and to lines which were called "mechanical"[48] by Descartes, were excluded by the latter in his calculus. But now by the calculus of Leibniz the whole of geometry is subjected to analytical computation, and those transcendent lines that Descartes called mechanical are also reduced to equations chosen to suit them, by considering the differences dx, ddx, etc., and the sums that are the inverses of these differences, as functions of the x's; and this, by merely introducing the calculus, whereas before this no other functions were admissible but x, xx, x^3, $\sqrt{x}$, etc., that is to say, powers and roots.[49] Hence it is easy to see that those who expressed these differences by 0, as did Fermat, Descartes, and even that rival, in his *Principia* published in 16—,[50] were by that very fact an extremely long way off from the differential calculus; for in this way neither gradation of the differences nor the differential functions of the several quantities can possibly be made out.

There does not exist anywhere the slightest trace of these methods having been practised by any one before Leibniz.[51] With

[48] Apollonian geometry comprised the conic sections or curves of the second degree according to Cartesian geometry; curves of a higher degree and of a transcendent nature, like the spiral of Archimedes, were included under the term "mechanical."

[49] The great discovery of Descartes was not simply the application of geometry; that had been done in simple cases ages before. Descartes recognized the principle that every property of the curve was included in its equation, if only it could be brought out. Thus Leibniz's greatest achievement was the recognition that the differential coefficients were also functions of the abscissa. The word function was applied to certain straight lines dependent on the curve, such as the abscissa itself, the ordinate, the chord, the tangent, the perpendicular, and a number of others (Cantor, III, preface, p. v). This definition is from a letter to Huygens in 1694. There is therefore a great advance made by 1714, the date of the *Historia*, since here it is at least strongly hinted that Leibniz has the algebraical idea of a function.

[50] With regard to Newton, at least, this is untrue. Without a direct reference to the original manuscript of Newton it is quite impossible to state whether even Newton *wrote* 0 or *o*; even then there may be a difficulty in deciding, for Gerhardt and Weissenborn have an argument over the matter, while Reiff prints it as 0. However this may be there is no doubt that Newton considered it as an infinitely small *unit of time*, only to be put equal to zero when it occurred as a factor of terms in an expression in which there also occurred terms that did not contain an infinitesimally small factor. This was bound to be the case, since Newton's $\dot{x}$ and $\dot{y}$ were velocities. In short, expressing Newton's notation in that of Leibniz, we have

$$\dot{x}o \text{ or } \dot{x}0 = (dx/dt) . dt$$

and therefore $\dot{x}o$ is an infinitesimal or a differential equal to Leibniz's dx.

[51] This is in a restricted sense true. No one seems to have felt the need of a second differentiation of an original function; those, who did, differen-

precisely the same amount of justice as his opponents display in now assigning such discoveries to Newton, any one could equally well assign the geometry of Descartes to Apollonius, who, although he possessed the essential idea of the calculus, yet did not possess the calculus.

For this reason also the new discoveries that were made by the help of the differential calculus were hidden from the followers of Newton's method, nor could they produce anything of real value nor even avoid inaccuracies until they learned the calculus of Leibniz, as is found in the investigation of the catenary as made by David Gregory.[52] But these contentious persons have dared to misuse the name of the English Royal Society, which body took pains to have it made known that no really definite decision was come to by them; and this is only what is worthy of their reputation for fair dealing, in that one of the two parties was not heard, indeed my friend himself did not know that the Royal Society had undertaken an inquiry into the matter. Else the names of those to whom it had entrusted the report would have been communicated to him,[53] so that they might either be objected to, or equipped for their task. He indeed, astounded not by their arguments but by the fictions that pervaded their attack on his good faith, considered such things unworthy of a reply, knowing as he did that it would be useless to defend his case before those who were unacquainted with this subject (i. e., the great majority of readers); also feeling that those who were skilled in the matter under discussion would readily perceive the injustice of the charge.[54] To this was added the reason that he was absent from home when these reports were circulated by his opponents, and returning home after an interval of two years and being occupied with other busi-

tiated once, and then worked upon the function thus obtained a second time in the same manner as in the first case. Barrow indeed considered only curves of continuous curvature, and the tangents to these curves; but Newton has the notation $\ddot{x}$, etc. But the idea had been used by Slusius in his *Mesolabum* (1659), where a general method of determining points of inflection is made to depend on finding the maximum and minimum values of the subtangent. Lastly, it can hardly be said that Leibniz's interpretation of $\int\int$ ever attained to the dignity of a double integral in his hands.

[52] David Gregory is not the only sinner! Leibniz, using his calculus, makes a blunder over osculations, and will not stand being told about it; he simply repeats in answer that he is right (Rouse Ball's *Short History*).

[53] The names of the committee were not even published with their report. In fact the complete list was not made public until De Morgan investigated the matter in 1852! For their names see De Morgan's *Newton*, p. 27.

[54] What then made Leibniz change his mind?

ness, it was then too late to find and consult the remains of his own past correspondence from which he might refresh his memory about matters that had happened so long ago as forty years previously. For transcripts of very many of the letters once written by him had not been kept; besides those that Wallis found in England and published with his consent in the third volume of his works, Leibniz himself had not very many.

Nevertheless, he did not lack for friends to look after his fair name; and indeed a certain mathematician, one of the first rank of our time[55] well skilled in this branch of learning and perfectly unbiased, whose good-will the opposite party had tried in vain to obtain, plainly stated, giving reasons of his own finding, and let it be known, not altogether with strict justice, that he considered that not only had that rival not invented the calculus, but that in addition he did not understand it to any great extent.[56] Another friend of the inventor[57] published these and other things as well in a short pamphlet, in order to check their base contentions. However it was of greater service to make known the manner and reasoning by which the discoverer arrived at this new kind of calculus; for this indeed has been unknown up till now, even to those perchance, who would like to share in this discovery. Indeed he himself had decided to explain it, and to give an account of the course of his researches in analysis partly from memory and partly from extant writings and remains of old manuscripts, and in this manner to illustrate in due form in a little book the history of this higher learning and the method of its discovery. But since at the time this was found to be impossible owing to the necessities of other business, he allowed this short statement of part of what there was to tell upon the matter to be published in the meantime by a friend who knew all about it,[58] so that in some measure public curiosity should be satisfied.

[55] It is established that this was Johann (John) Bernoulli; see Cantor, III, p. 313f; Gerhardt gives a reference to Bossut's *Geschichte,* Part II, p. 219.

[56] This seems to be an intentional misquotation from Bernoulli's letter, which stated that Newton did not understand the meaning of higher differentiations. At least, that is what Cantor says was given in the pamphlet.

[57] It is established that the pamphlet referred to was also an anonymous contribution by Leibniz himself! Is it strange that hard things are both thought and said of such a man?

[58] Again this is Leibniz himself! Had he then no friends at all to speak for him and dare subscribe their signatures to the opinion? Unfortunately Tschirnhaus was dead at the time of the publication of the *Commercium*

The author of this new analysis, in the first flower of his youth, added to the study of history and jurisprudence other more profound reflections for which he had a natural inclination. Among the latter he took a keen delight in the properties and combinations of numbers; indeed, in 1666 he published an essay, *De Arte Combinatoria,* afterward reprinted without his sanction. Also, while still a boy, when studying logic he perceived that the ultimate analysis of truths that depended on reasoning reduced to two things, definitions and identical truths, and that these alone of the essentials were primitive and undemonstrable. When it was stated in contradiction that identical truths were useless and nugatory, he gave illustrative proofs to the contrary. Among these he gave a demonstration that that mighty axiom, "The whole is greater than its part," could be proved by a syllogism of which the major term was

Epistolicum, but he could have spoken with overwhelming authority, as Leibniz's co-worker in Paris, at any time between the date of Leibniz's review of Newton's *De Quadratura* in the *Acta Eruditorum* until his death in 1708, even if he had died before the publication of Keill's attack in the *Phil. Trans.* of that year was made known to him. Does not this silence on the part of Tschirnhaus, the personal friend of Leibniz, rather tend to make Leibniz's plea, that his opponents had had the shrewdness to wait till Tschirnhaus, among others, was dead, recoil on his own head, in that he has done the very same thing? Leibniz must have known the feeling that this review aroused in England, and, Huygens being dead, Tschirnhaus was his only reliable witness. Of course I am not arguing that Leibniz did found his calculus on that of Newton. I am fully convinced that they both were indebted to Barrow, Newton being so even more than Leibniz, and that they were perfectly independent of one another in the development of the *analytical* calculus. Newton, with his great knowledge of and inclination toward geometrical reasoning, backed with his personal intercourse with Barrow, could appreciate the finality of Barrow's proofs of the differentiation of a product, quotient, power, root, logarithm and exponential, and the trigonometrical functions, in a way that Leibniz could not. But Newton never seems to have been accused of plagiarism from Barrow; even if he had been so accused, he probably had ready as an answer, that Barrow had given him permission to make any use he liked of the instruction that he obtained from him. Leibniz, when so accused, replied by asserting, through confusion of memory I suggest, that he got his first idea from the works of Pascal. Each developed the germ so obtained in his own peculiar way; Newton only so far as he required it for what he considered his main work, using a notation that was of greatest convenience to him, and finally falling back on geometry to provide himself with what appealed to him as rigorous proof; Leibniz, more fortunate in his philosophical training and his lifelong effort after symbolism, has ready to hand a notation, almost developed and perfected when applied to finite quantities, which he saw with the eye of genius could be employed as usefully for infinitesimals. De Morgan justly remarks that one dare not accuse either of these great men of deliberate untruth with regard to specific facts; but it must be admitted that neither of them can be considered as perfectly straightforward; and the political similitude, which Cantor speaks of, in which nothing is too bad to be said of an opponent, seems to have applied just as much to the mathematician of the day as to the politician.

a definition and the minor term an identity.[59] For if one of two things is equal to a part of another the former is called the less, and the later the greater; and this is to be taken as the definition. Now, if to this definition there be added the following identical and undemonstrable axiom, "Every thing possessed of magnitude is equal to itself," i. e., $A = A$, then we have the syllogism:

Whatever is equal to a part of another, is less than that other:
(by the definition)

But the part is equal to a part of the whole:
(i. e., to itself, by identity)

Hence the part is less than the whole.   Q. E. D.

As an immediate consequence of this he observed that from the identity $A = A$, or at any rate from its equivalent, $A - A = 0$, as may be seen at a glance by straightforward reduction, the following very pretty property of differences arises, namely:

$$A \quad \underbrace{-A+B}_{+\ L} \quad \underbrace{-B+C}_{+\ M} \quad \underbrace{-C+D}_{+\ N} \quad \underbrace{-D+E}_{+\ P} \quad -E \ = \ 0$$

If now A, B, C, D, E are supposed to be quantities that continually increase in magnitude, and the differences between successive terms are denoted by L, M, N, P, it will then follow that

$$A + L + M + N + P - E = 0,$$
$$\text{i. e.,} \qquad L + M + N + P = E - A;$$

that is, the sums of the differences between successive terms, no matter how great their number, will be equal to the difference

[59] This was given in more detail in the first draught of this essay (G. 1846, p. 26): Hitherto, while still a pupil, he kept trying to reduce logic itself to the same state of certainty as arithmetic. He perceived that occasionally from the first figure there could be derived a second and even a third, without employing conversions (which themselves seemed to him to be in need of demonstration), but by the sole use of the principle of contradiction. Moreover, these very conversions could be proved by the help of the second and third figures, by employing theorems of identity; and then now that the conversion had been proved, it was possible to prove a fourth figure also by its help, and this latter was thus more indirect than the former figures. He marveled very much at the power of identical truths, for they were generally considered to be useless and nugatory. But later he considered that the whole of arithmetic and geometry arose from identical truths, and in general that all undemonstrable truths depending on reasoning were identical, and that these combined with definitions yield identical truths. He gave as an elegant example of this analysis a proof of the theorem, The whole is greater than its part.

between the terms at the beginning and the end of the series.[60]
For example, in place of A, B, C, D, E, let us take the squares,
0, 1, 4, 9, 16, 25, and instead of the differences given above, the
odd numbers, 1, 3, 5, 7, 9, will be disclosed; thus

$$0 \quad 1 \quad 4 \quad 9 \quad 16 \quad 25$$
$$1 \quad 3 \quad 5 \quad 7 \quad 9$$

From which is evident that

$$1 + 3 + 5 + 7 + 9 = 25 - 0 = 25,$$
$$\text{and} \quad 3 + 5 + 7 + 9 = 25 - 1 = 24;$$

and the same will hold good whatever the number of terms or the
differences may be, or whatever numbers are taken as the first and
last terms. Delighted by this easy, elegant theorem, our young friend
considered a large number of numerical series, and also proceeded
to the second differences or differences of the differences,[61] the

[60] It is fairly certain that Leibniz could not possibly at this time have
perceived that in this theorem he has the germ of an integral. The path to
the higher calculus lay through geometry. As soon as Leibniz attained to a
sufficient knowledge of this subject he would recognize the area under a curve
between a fixed ordinate and a variable one as a set of magnitudes of the
kind considered, the ordinates themselves being the differences of the set; he
would see that there was no restriction on the number of steps by which the
area attained its final size. Hence, in this theorem he has a proof to hand
that integration as a determination of an area is the inverse of a difference.
This does not mean the inverse of a differentiation, i. e., the determination
of a *rate*, or the drawing of a tangent. As far as I can see, Leibniz was far
behind Newton in this, since Newton's fluxions were founded on the idea of
a rate; also Leibniz apparently does not demonstrate the rigor of a method of
infinitely narrow rectangles.

[61] It is a pity that we are not told the date at which Leibniz read his
Wallis; it is a greater pity that Gerhardt did not look for a Wallis in the
Hanover Library and see whether it had the date of purchase on it (for I
have handled lately several of the books of this time, and in nearly every
case I found inserted on the title page the name of the purchaser and the date
of purchase). I make this remark, because there arises a rather interesting
point. Wallis, in his *Arithmetica Infinitorum*, takes as the first term of all his
series the number 0, and in one case he mentions that the differences of the differences of the cubes is an arithmetical series. He also works out fully the sums
of the figurate numbers (or as Leibniz calls them the combinatory numbers);
the general formulas for these sums he calls their *characteristics*. He also remarks on the fact that any number (see table, p. 32) can be obtained by the addition of the one before it and the one above it (which is itself the sum of all the
numbers in the preceding column above the one to the left of that which he
wishes to obtain). Thus, in the fourth column 4 is the sum of 3 (to the left)
and I (above), i. e., the sum of the two first numbers in column three; 10
is the sum of 6 (to the left) and 4 (above, which has been shown to be the
sum of the first two numbers of column three), and therefore 10 is the sum
of the first three numbers in column three. Now my point is, assuming it to
have been impossible that Leibniz had read Wallis at the time that he was
compiling his *De Arte,* we have here another example, free from all suspicion,
of that series of instances of independent contemporary discoveries that seems
to have dogged Leibniz's career.

third differences or the differences between the differences of the differences, and so on. He also observed that for the natural numbers, i. e., the numbers in order proceeding from 0, the second differences vanished, as also did the third differences for the squares, the fourth differences for the cubes, the fifth for the biquadrates, the sixth for the surdesolids,[62] and so on; also that the first differences for the natural numbers were constant and equal to 1; the second differences for the square, 1.2, or 2; the third for the cubes, 1.2.3, or 6; the fourth for the biquadrates, 1.2.3.4, or 24; the fifth for the surdesolids, 1.2.3.4.5, or 120, and so on. These things it is admitted had been previously noted by others, but they were new to him, and by their easiness and elegance were in themselves an inducement to further advances. But especially he considered what he called "combinatory numbers," such as are usually tabulated as in the margin. Here a preceding series, either horizontal or vertical, always contains the first differences of the series immediately following it, the second differences of the one next after that, the third differences of the third, and so on. Also, each series, either horizontal or vertical contains the sums of the series immediately preceding it, the sums of the sums or the second sums of the series next before that, the third sums of the third, and so on. But, to give something not yet common knowledge, he also brought to light certain general theorems on differences and sums, such as the following. In the series, *a, b, c, d, e*, etc., where the terms continually decrease without limit we have

1	1	1	1	1	1
1	2	3	4	5	6
1	3	6	10	15	21
1	4	10	20	35	56
1	5	15	35	70	126
1	6	21	56	126	252
1	7	28	84	210	462

Terms	*a*	*b*	*c*	*d*	*e*	etc.
1st diff.		*f*	*g*	*h*	*i*	*k* etc.
2nd diff.		*l*	*m*	*n*	*o*	*p* etc.
3rd diff.			*q*	*r*	*s*	*t* *u* etc.
4th diff.			*β*	*γ*	*δ*	*ε* *θ* etc.
etc.			*γ*	*μ*	*ν*	*ρ* *υ* etc.

---

[62] The name surdesolid to denote the fifth power is used by Oughtred, according to Wallis. By Cantor the invention of the term seems to be credited to Dechales, who says, "The fifth number from unity is called by some people the quadrato-cubus, but this is ill-done, since it is neither a square nor a cube and cannot thus be called the square of a cube nor the cube of a square: we shall call it supersolidus or surde solidus" (Cantor, III, p. 16). Wallis himself uses "sursolid."

Taking a as the first term, and ω as the last, he found

$$a - \omega = 1f + 1g + \;\;1h + \;\;1i + \;\;1k + \text{etc.,}$$
$$a - \omega = 1l + 2m + \;\;3n + \;\;4o + \;\;5p + \text{etc.,}$$
$$a - \omega = 1q + 3r + \;\;6s + 10t + 15u + \text{etc.,}$$
$$a - \omega = 1\beta + 4\gamma + 10\delta + 20\epsilon + 35\theta + \text{etc.,}$$
etc.

Again we have[63]

$$a - \omega = \begin{cases} + 1f \\ + 1f - 1l \\ + 1f - 2l + 1q \\ + 1f - 3l + 3q - 1\beta \\ + 1f - 4l + 6q - 4\beta + 1\lambda \\ \quad \text{etc.} \quad \text{etc.} \quad \text{etc,} \end{cases}$$

Hence, adopting a notation invented by him at a later date, and denoting any term of the series generally by y (in which case $a = y$ as well), we may call the first difference dy, the second ddy, the third d^3y, the fourth d^4y; and calling any term of another of the series x, we may denote the sum of its terms by $\int x$, the sum of their sums or their second sum by $\int\int x$, the third sum by $\int^3 x$, and the fourth sum by $\int^4 x$. Hence, supposing that

$$1 + 1 + 1 + 1 + 1 + \text{etc.} = x,$$

or that x represents the natural numbers, for which $dx = 1$, then

---

[63] This theorem is one of the fundamental theorems in the theory of the summation of series by finite differences, namely,

$$\Delta^m u_n = u_{n+m} - {}_mC_1 \cdot u_{n+m-1} + {}_mC_2 \cdot u_{n+m-2} - \text{etc.,}$$

which is usually called the direct fundamental theorem; for although Leibniz could not have expressed his results in this form since he did not know the sums of the figurate numbers as generalized formulas (or I suppose not, if he had not read Wallis), and apparently his is only a special case, yet it must be remembered that any term of the first series can be chosen as the first term. It is interesting to note that the second fundamental theorem, the inverse fundamental theorem, was given by Newton in the *Principia,* Book III, lemma V, as a preliminary to the discussion on comets at the end of this book. Here he states the result, without proof, as an interpolation formula; (it is frequently referred to as Newton's Interpolation Formula); it may however be used as an extrapolation formula, in which case we have

$$u_{m+n} = u_m + {}_nC_1 \cdot \Delta u_m + {}_nC_2 \cdot \Delta^2 u_m + \text{etc.}$$

In the two formulas as given here, the series are

$$u_1 \qquad u_2 \qquad u_3 \qquad u_4 \qquad u_5 \text{ etc.}$$
$$\Delta u_1 \quad \Delta u_2 \quad \Delta u_3 \quad \Delta u_4 \qquad \text{etc.}$$
$$\Delta^2 u_2 \quad \Delta^2 u_3 \quad \Delta^2 u_4 \qquad \text{etc.} \quad \text{and so on.}$$

$$1 + 3 + \; 6 + 10 + \text{etc.} = \int x,$$
$$1 + 4 + 10 + 20 + \text{etc.} = \int \int x,$$
$$1 + 5 + 15 + 35 + \text{etc.} = \int^3 x,$$

and so on. Finally it follows that

$$y - \omega = dy \cdot x - ddy \cdot \int x + d^3y \cdot \int \int x - d^4y \cdot \int^3 x + \text{etc.} \,;$$

and this is equal to y, if we suppose that the series is continued to infinity, or that ω becomes zero. Hence also follows the sum of the series itself, and we have

$$\int y = yx - dy \cdot \int x + ddy \cdot \int \int x - d^3y \cdot \int^3 x + \text{etc.}^{64}$$

These two like theorems possess the uncommon property that they are equally true in either differential calculus, the numerical or the infinitesimal; of the distinction between them we will speak later.[65]

[64] What are we to understand by the inclusion of this series in this connection? Does Leibniz intend to claim this as his? I have always understood that this is due to Johann Bernoulli, who gave it in the *Acta Eruditorum* for 1694, in a slightly different form, and proved by direct differentiation; and that Brook Taylor obtained it as a particular case of a general theorem in and by *finite differences*. If Leibniz intended to claim it, he has clearly anticipated Taylor. It is quite possible that Leibniz had done so, even in his early days; and as soon as in 1675, or thereabouts, he had got his signs for differentiation and integration, it is possible that he returned to this result and expressed it in the new notation; for the theorem follows so perfectly naturally from the last expression given for $a - \omega$. But it is hardly probable, for Leibniz would almost certainly have shown it to Huygens and mentioned it.

The other alternative is that here he is showing how easily Bernoulli's series could have been found in a much more *general form*, i. e., as a theorem that is true (as he indeed states) for finite differences as well as for infinitesimals; the inclusion of this statement makes it very probable that this supposition is a correct one. This leads to a pertinent, or impertinent, question. Brook Taylor's *Methodus Incrementorum* was published in 1715; the *Historia* was written some time between 1714 and 1716; Gerhardt states that there were two draughts of the latter, and that he is giving the second of these. In justice to Leibniz there should be made a fresh examination of the two draughts, for if this theorem is not given in the original draught it lays Leibniz open to further charge of plagiarism. I fully believe that the theorem will be found in the first draught as well and that my alternative suggestion is the correct one.

In any case, the tale of the *Historia* is confused by the interpolation of the symbolism invented later (as Leibniz is careful to point out). The question is whether this was not intentional. And this query is not impertinent, considering the manner in which Leibniz refrains from giving dates, or when we compare the essay in the *Acta Eruditorum*, in which he gives to the world the description of his method. Weissenborn considers that "this is not adapted to give an insight into his methods, and it certainly looks as if Leibniz wished deliberately to prevent this." Cf. Newton's "anagram" (sic), and the Geometry of Descartes, for parallels.

[65] In reference to the employment of the calculus to diagrammatic geometry, as will be seen later, Leibniz says:
"But our young friend quickly observed that the differential calculus could be employed with figures in an even more wonderfully simple manner

However, the application of numerical truths to geometry, as well as the consideration of infinite series, was at that time at all events unknown to our young friend, and he was content with the satisfaction of having observed such things in series of numbers. Nor did he then, except for the most ordinary practical rules, know anything about geometry;[66] he had scarcely even considered Euclid with anything like proper attention, being fully occupied with other studies. However, by chance he came across the delightful contemplation of curves by Leotaud, in which the author deals with the quadrature of lunules, and Cavalieri's geometry of indivisibles;[67] having given these some slight consideration, he was delighted with the facility of their methods. However, at the time he was in no mind to go fully into these more profound parts of mathematics; although just afterwards he gave attention to the study of physics and practical mechanics, as may be understood from his essay that he published on the *Hypothesis of Physics*.[68]

He then became a member of the Revision Council[69] of the Most Noble the Elector of Mainz; later, having obtained permission from this Most Gracious and Puissant Prince (for he had taken our young friend into his personal service when he was about to

than it was with numbers, because *with figures the differences were not comparable with the things which differed*; and as often as they were connected together by addition or subtraction, being incomparable with one another, the less vanished in comparison with the greater."

[66] This makes what has just gone before date from the time previous to his reading of the work of Cavalieri. See note following.

[67] This is about the first place in which it is possible to deduce an exact date, or one more or less exact. According to Leibniz's words that immediately follow it may be deduced that it was somewhere about twelve months before the publication of the Hypothesis of Physics—if we allow for a slight interval between the dropping of the geometry and the consideration of the principles of physics and mechanics, and a somewhat longer interval in which to get together the ideas and materials for his essay—that he had finished his "slight consideration" of Leotaud and Cavalieri. This would make the date 1670, and his age 24.

[68] This essay founded the explanation of all natural phenomena on motion, which in turn was to be explained by the presence of an all-pervading ether; this ether constituted light.

[69] The dedication of the *Nova methodus* in 1667 to the Elector of Mainz (ancient name Moguntiacum) procured for Leibniz his appointment in the service of the latter, first as an assistant in the revision of the statute-book, and later on the more personal service of maintaining the policy of the Elector, that of defending the integrity of the German Empire against the intrigues of France, Turkey and Russia, by his pen.

leave[70] and go further afield) to continue his travels, he set out for Paris in the year 1672. There he became acquainted with that genius, Christiaan Huygens, to whose example and precepts he always declared that he owed his introduction to higher mathematics. At that time it so happened that Huygens was engaged on his work with regard to the pendulum. When Huygens brought our young friend a copy of this work as a present and in the course of conversation discussed the nature of the center of gravity, which our young friend did not know very much about, the former explained to him shortly what sort of thing it was and how it could be investigated.[71] This roused our young friend from his lethargy, for he looked upon it as something of a disgrace that he should be ignorant of such matters.[72]

Now it was impossible for him to find time for such studies just then; for almost immediately, at the close of the year, he crossed the Channel to England in the suite of the envoy from Mainz, and stayed there for a few weeks with the envoy. Having been introduced by Henry Oldenburg, at that time secretary to the Royal Society, he was elected a member of that illustrious body. He did not however at that time discuss geometry with any one (in truth at that time he was quite one of the common herd as regards this subject); he did not on the other hand neglect chemistry, consulting that excellent man, Robert Boyle, on several occasions. He also came across Pell accidentally, and he described to him certain of his own observations on numbers; and Pell told him that they were not new, but that it had been recently made known by Nicolaus Mercator, in his *Hyperbolae Quadratura,* that the differences of the powers of the natural numbers, when taken continuously, finally vanished; this made Leibniz obtain the work of Nicolaus

[70] This probably refers to the time when his work on the statute-book was concluded, and Leibniz was preparing to look for employment elsewhere.

[71] This is worthy of remark, seeing that Leibniz had attempted to explain gravity in the *Hypothesis physica nova* by means of his concept of an ether. The conversation with Huygens had results that will be seen later in a manuscript (see § 4, p. 65) where Leibniz obtains quadratures *"ex Centrobarycis."* It also probably had a great deal to do with Leibniz's concept of a "moment."

[72] The use of the word *veterno*—which I have translated "lethargy" as being the nearest equivalent to the fundamental meaning, the sluggishness of old age—coupled with his remark that he was in no mind to enter fully into these more profound parts of mathematics, sheds a light upon the reason why he had so far done no geometry. Also the last words of the sentence give the stimulus that made him cast off this lethargy; namely, shame that he should appear to be ignorant of the matter. This would seem to be one of the great characteristics of Leibniz, and might account for much, when we come to consider the charges that are made against him.

Mercator.[73] At that time he did not become acquainted with Collins; and, although he conversed with Oldenburg on literary matters, on physics and mechanics, he did not exchange with him even one little word on higher geomery, much less on the series of Newton. Indeed, that he was almost a stranger to these subjects, except perhaps in the properties of numbers, even that he had not paid very much attention to them, is shown well enough by the letters which he exchanged with Oldenburg, which have been lately published by his opponents. The same fact will appear clearly from those which they say have been preserved in England; but they suppressed them,[74] I firmly believe, because it would be quite clear from them that up to then there had been no correspondence between him and Oldenburg on matters geometrical. Nevertheless, they would have it credited (not indeed with the slightest evidence brought forward in favor of the supposition) that certain results obtained by Collins, Gregory and Newton, which were in the possession of Oldenburg, were communicated by him to Leibniz.

On his return from England to France in the year 1673,[75] having meanwhile satisfactorily performed his work for the Most Noble Elector of Mainz, he still by his favor remained in the service of Mainz; but his time being left more free, at the instigation of Huygens he began to work at Cartesian analysis (which aforetime had been beyond him),[76] and in order to obtain an insight

[73] We have here a parallel (or a precedent) for my suggestion that Leibniz was mentally confusing Barrow and Pascal as the source of his inspiration for the characteristic triangle. For here, without any doubt whatever, is a like confusion. What Pell told him was that his theorems on numbers occurred in a book by Mouton entitled *De diametris apparentibus Solis et Lunae* (published in 1670). Leibniz, *to defend himself from a charge of plagiarism,* made haste to borrow a copy from Oldenburg and found to his relief that not only had Mouton got his results by a different method, but that his own were more general. The words in italics are interesting.

Of course these words are not italicized by Gerhardt, from whom this account has been taken (G. 1848, p. 19); nor does he remark on Leibniz's lapse of memory in this instance. Further there is no mention made of it in connection with the *Historia,* i. e., in G. 1846. Is it that Gerhardt, as counsel for the defense, is afraid of spoiling the credibility of his witness by proving that part of his evidence is unreliable? Or did he not become aware of the error till afterward? See Cantor, III, p. 76.

[74] An instance is referred to on p. 85 of De Morgan's *Newton,* showing the sort of thing that was done by the committee. This however is not connected with a letter to Oldenburg, but to Collins. It may be taken as a straw that shows the way the wind blew.

[75] Observe that nothing has been said of the fact that Leibniz had purchased a copy of Barrow and took it back with him to Paris.

[76] Cf. the remark in the postscript to Bernoulli's letter, where Leibniz says that the work of Descartes, looked at at about the same time as Clavius, that is, while he was still a youth, "seemed to be more intricate."

into the geometry of quadratures he consulted the *Synopsis Geo-metriae* of Honoratus Fabri, Gregory St. Vincent, and a little book by Dettonville (i. e., Pascal).[77] Later on from one example given by Dettonville, a light suddenly burst upon him, which strange to say Pascal himself had not perceived in it. For when he proves the theorem of Archimedes for measuring the surface of a sphere or parts of it, he used a method in which the whole surface of the solid formed by a rotation round any axis can be reduced to an equivalent plane figure. From it our young friend made out for himself the following general theorem.[78]

Portions of a straight line normal to a curve, intercepted between the curve and an axis, when taken in order and applied at right angles to the axis give rise to a figure equivalent to the moment of the curve about the axis.[79]

When he showed this to Huygens the latter praised him highly and confessed to him that by the help of this very theorem he had found the surface of parabolic conoids and others of the same sort, stated without proof many years before in his work on the pendulum clock. Our young friend, stimulated by this and pondering on the fertility of this point of view, since previously he had considered infinitely small things such as the intervals between the ordinates in the method of Cavalieri and such only, studied the triangle $_1Y\,D\,_2Y$, which he called the Characteristic Triangle,[80]

[77] The *libellus* referred to would seem to be the work on the cycloid, written by Pascal in the form of letters, from one Amos Dettonville, to M. de Carcavi.

[78] This theorem is given, and proved by the method of indivisibles, as Theorem I, of Lecture XII in Barrow's *Lectiones Geometricae*; and Theorem II is simply a corollary, in which it is remarked:
"Hence the surfaces of the sphere, both the spheroids, and the conoids receive measurement...."
The proof of these two theorems is given at the end of this section as a supplement. See also Note 46, for its significance.

[79] The whole context here affords suggestive corroboration in favor of the remarks made in Note 31 on the use of the word "moment," though the connection with the determination of the center of gravity is here over-shadowed by its connection with the surface formed by the rotation of an arc about an axis.

[80] The figure given is exactly that given by Gerhardt, with the unimportant exception that, for convenience in printing, I have used U instead of Gerhardt's Θ, a V instead of his ח (a Hebrew T), and a Q for his Π. I take it, of course, that Gerhardt's diagram is an exact transcript of Leibniz's, and it is interesting to remark that Leibniz seems to be endeavoring to use T's for all points on the tangent, and P's for points on the normal, or perpendicular, as it is rendered in the Latin.
This diagram should be compared with that in the "postscript" written nine or ten years before. Note the complicated diagram that is given here.

whose sides D_1Y, D_2Y are respectively equal to $_1X_2X$, $_1Z_2Z$,[81] parts of the coordinates or coabscissae AX, AZ, and its third side $_1Y_2Y$ a part of the tangent TV, produced if necessary.

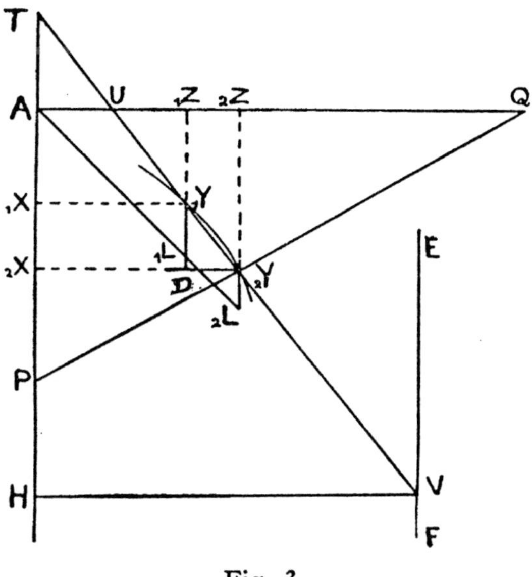

Fig. 3.

Even though this triangle is indefinite (being infinitely small), yet he perceived that it was always possible to find definite triangles similar to it. For, suppose that AXX, AZZ are two straight lines at right angles, and AX, AZ the coabscissae, YX, YZ the coordinates, TUV the tangent, PYQ the perpendicular, XT, ZU the subtangents, XP, ZQ the subnormals; and lastly let EF be drawn

and the introduction of the secant that is ultimately the tangent, which does not appear in the first figure. From what follows, this is evidently done in order to introduce the further remarks on the similar triangles. It adds to the confusion when an effort is made to determine the dates at which the several parts were made out. For instance, the remark that finite triangles can be found similar to the characteristic triangle probably belongs approximately to the date of his reply to the assertions of Nieuwentijt, which will be referred to later.

[81] The notation introduced in the lettering should be remarked. His early manuscripts follow the usual method of the time in denoting different positions of a variable line by the same letter, as in Wallis and Barrow, though even then he is more consistent than either of the latter. He soon perceives the inconvenience of this method, though as a means of generalizing theorems it has certain advantages. We therefore find the notation C, (C), ((C)), for three consecutive points on a curve, as occurs in a manuscript dated (or it should be) 1675. This notation he is still using in 1703; but in 1714, he employs a subscript prefix. This is all part and parcel with his usual desire to standardize and simplify notations.

parallel to the axis AX; let the tangent TY meet EF in V, and from V draw VH perpendicular to the axis. Then the triangles $_1YD_2Y$, TXY, YZU, TAU, YXP, QZY, QAP, THV, and as many more of the sort as you like, are all similar. For example, from the similar triangles $_1YD_2Y$, $_2Y_2XP$, we have $P_2Y._1YD = _2Y_2X._2Y_1Y$; that is, the rectangle contained by the perpendicular P_2Y and $_1YD$ (or the element of the axis, $_1X_2X$) is equal to the rectangle contained by the ordinate $_2Y_2X$ and the element of the curve, $_1Y_2Y$, that is, to the moment of the element of the curve about the axis. Hence the whole moment of the curve is obtained by forming the sum of these perpendiculars to the axis.

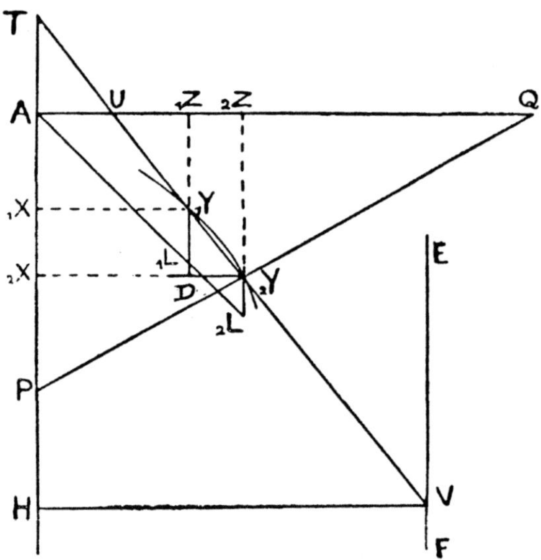

Also, on account of the similar triangles $_1YD_2Y$, THV, we have $_1Y_2Y:_2YD=TV:VH$, or $VH._1Y_2Y=TV._2YD$; that is, the rectangle contained by the constant length VH and the element of the curve, $_1Y_2Y$, is equal to the rectangle contained by TV and $_2YD$, or the element of the coabscissa, $_1Z_2Z$. Hence the plane figure produced by applying the lines TV in order at right angles to AZ is equal to the rectangle contained by the curve when straightened out and the constant length HV.

Again, from the similar triangles $_1YD_2Y$, $_2Y_2XP$, we have $_1YD:D_2Y=_2Y_2X:_2XP$, and thus $_2XP._1YD=_2Y_2X.D_2Y$, or the sum of the subnormals $_2XP$, taken in order and applied to the axis, either to $_1YD$ or to $_1X_2X$, will be equal to the sum of the products of the ordinates $_2Y_2X$ and their elements, $_2YD$, taken in order. But straight lines that continually increase from zero, when each is multiplied by its element of increase, form altogether a triangle. Let then AZ always be equal to ZL, then we get the right-angled triangle AZL, which is half the square on AZ; and thus the figure that is produced by taking the subnormals in order and applying them perpendicular to the axis will be always equal to half the square on the ordinate. Thus, to find the area of a given figure, another figure is sought such that its subnormals are respectively equal to the ordinates of the given figure, and then this second figure is the quadratrix of the given one; and thus from this extremely elegant consideration we obtain the reduction of the areas of surfaces described by rotation[82] to plane quadratures, as well as the rectification of curves; at the same time we can reduce these quadratures of figures to an inverse problem of tangents. From these results,[83] our young friend wrote down a large collection of theorems (among which in truth there were many that were lacking in elegance) of two kinds. For in some of them only definite magnitudes were dealt with, after the manner not only of Cavalieri, Fermat, Honoratus Fabri, but also of Gregory St. Vincent, Guldinus, and Dettonville; others truly depended on infinitely small magnitudes, and advanced to a much greater extent. But later our young friend did not not trouble to go on with these matters, when he noticed that the same method had been brought into use and perfected by not only Huygens, Wallis, Van Huraet, and Neil, but also by James Gregory and Barrow. However it did not seem to me to be altogether useless to explain at this juncture, as is plain from what I have given,[84] the steps by which he attained to greater things, and also the manner in which, as if led by the hand, those who are at present but beginners[85] with regard

[82] This sentence conclusively proves that Leibniz's use of the moment was for the purposes of quadrature of surfaces of rotation.

[83] "From these results"—which I have suggested he got from Barrow—"our young friend wrote down a large collection of theorems." These theorems Leibniz probably refers to when he says that he found them all to have been anticipated by Barrow, "when his Lectures appeared." I suggest that the "results" were all that he got from Barrow on his first reading, and that the "collection of theorems" were found to have been given in Barrow when Leibniz referred to the book again, after his geometrical knowledge was improved so far that he could appreciate it.

[84] The use of the first person is due to me. The original is impersonal, but is evidently intended by Leibniz to be taken as a remark of the writer, "the friend who knew all about it." The distinction is marked better by the use of the first personal pronoun than in any other way.

[85] Query, all except Leibniz, the Bernoullis, and one or two others.

to the more abstruse parts of geometry may hope to rise to greater heights.

Now Leibniz worked these things out at Paris in the year 1673 and part of 1674. But in the year 1674 (so much it is possible to state definitely), he came upon the well-known arithmetical tetragonism;[86] and it will be worth while to explain how this was accomplished. He once happened to have occasion to break up an area into triangles formed by a number of straight lines meeting in a

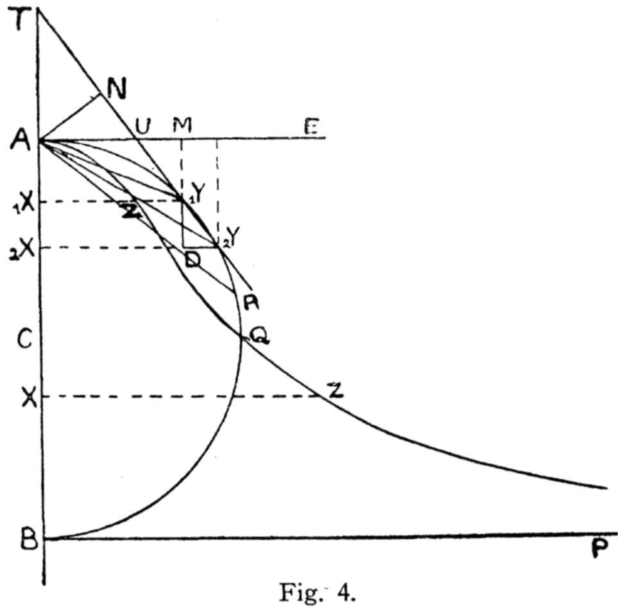

Fig. 4.

point, and he perceived that something new could be readily obtained from it.[87]

In Fig. 4, let any number of straight lines, AY, be drawn to the curve AYR, and let any axis AC be drawn, and AE, a normal or coaxis to it; and let the tangent at Y to the curve cut them in T and U. From A draw AN perpendicular to the tangent; then

[86] Tetragonism = quadrature; the arithmetical tetragonism is therefore Leibniz's value for π as an infinite series, namely,

"The area of a circle, of which the square on the diameter is equal to unity, is given by the series

$$\frac{1}{1} - \frac{1}{3} + \frac{1}{5} - \frac{1}{7} + \frac{1}{9} - \frac{1}{11} + \text{etc."}$$

[87] This is clearly original as far as Leibniz is concerned; but the consideration of a polar diagram is to be found in many places in Barrow. Barrow however forms the polar differential triangle, as at the present time, and does not use the rectangular coordinate differential triangle with a polar figure; nor does Wallis. We see therefore that Leibniz, as soon as ever he follows his own original line of thinking, immediately produces something good.

it is plain that the elementary triangle A_1Y_2Y is equal to half the rectangle contained by the element of the curve $_1Y_2Y$ and AN. Now draw the characteristic triangle mentioned above, $_1YD_2Y$, of which the hypotenuse is a portion of the tangent or the element of the arc, and the sides are parallel to the axis and the coaxis. It is then plain from the similar triangles ANU, $_1YD_2Y$, that $_1Y_2Y : _1YD = AU : AN$, or $AU . _1YD$ or $AU . _1X_2X$ is equal to $AN . _1Y_2Y$, and this, as has been already shown, is equal to double the triangle A_1Y_2Y. Thus if every AU is supposed to be transferred to XY, and taken in it as AZ,[88] then the trilinear space AXZA so formed will be equal to twice the segment $AY\smile A$,[89] included between the straight line AY and the arc AY. In this way are obtained what he called the figures of segments or the proportionals of a segment. A similar method holds good for the case in which the point is not taken on the curve, and in this manner he obtained the proportional trilinear figures for sectors cut off by lines meeting in the point; and even when the straight lines had their extremities not in a line but in a curve (which one after the other they touched), none the less on that account were useful theorems made out.[90] But this is not a fit occasion to follow out such matters; it is sufficient for our purpose to consider the figures of segments, and that too only for the circle. In this case, if the point A is taken at the beginning of the quadrant AYQ, the curve AZQZ will cut the circle at Q, the other end of the quadrant, and thence descending will be asymptotic to the base BP (drawn at right angles to the diameter at its other end B); and, although extending to infinity, the whole

---

[88] This is evidently a misprint; it is however curious that it is repeated in the second line of the next paragraph. Probably, therefore, it is a misreading due to Gerhardt, who mistakes AZ for the letters XZ, as they ought to be; and has either not verified them from the diagram, or has refrained from making any alteration.

[89] The symbol $\smile$ is here to be read as "and then along the arc to."

[90] Probably refers to Leibniz's work on curvature, osculating circles, and evolutes, as given in the *Acta Eruditorum* for 1686, 1692, 1694. It is to be noted that with Leibniz and his followers the term evolute has its present meaning, and as, such was first considered by Huygens in connection with the cycloid and the pendulum. It signified something totally different in the work of Barrow, Wallis and Gregory. With them, if the feet of the ordinates of a curve are, as it were, all bunched together in a point, so as to become the *radii vectores* of another curve, without rupturing the curve more than to alter its curvature (the area being thus halved), then the first curve was called the evolute of the second and the second the involute of the first. See Barrow's *Lectiones Geometricae*, Lecture XII, App. III, Prob. 9, and Wallis's *Arithmetica Infinitorum*, where it is shown that the evolute, in this sense, of a parabola is a spiral of Archimedes.

figure, included between the diameter AB, the base BP...., and the curve AZQZ.... asymptotic to it, will be equal to the circle on AB as diameter.

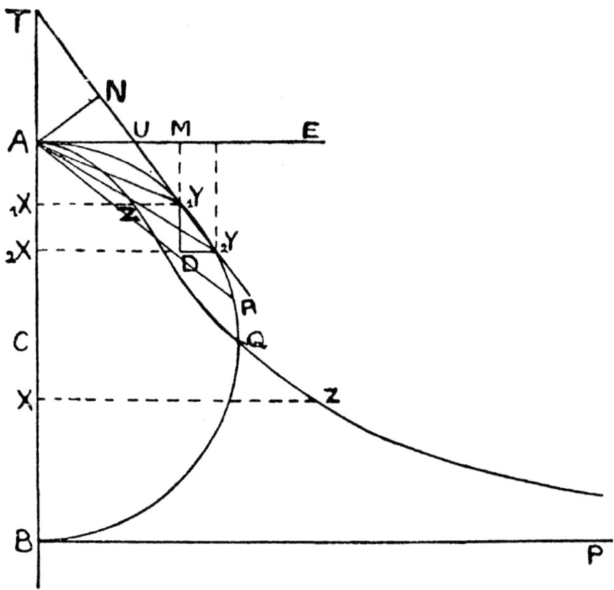

But to come to the matter under discussion, take the radius as unity, put AX or UZ = x, and AU or AZ = z, then we have $x = 2zz :, 1 + zz$;[91] and the sum of all the x's applied to AU, which at the present time we call $\int x\, dz$, is the trilinear figure AUZA, which is the complement of the trilinear figure AXZA, and this has been shown to be double the circular segment.

The author obtained the same result by the method of transmutations, of which he sent an account to England.[92] It is required to form the sum of all the ordinates $\sqrt{(1-xx)} = y$; suppose $y = \pm 1 \mp xz$, from which $x = 2z :, 1 + zz$, and $y = \pm zz \mp 1, :, zz + 1$; and thus again all that remains to be done is the summation of rationals.

This seemed to him to be a new and elegant method, as it did to Newton also, but it must be acknowledged that it is not

____

[91] The colon is used as a sign of division, and the comma has the significance of a bracket for all that follows. It is curious to notice that Leibniz still adheres to the use of xx for x^2, while he uses the index notation for all the higher powers, just as Barrow did; also, that the bracket is used under the sign for a square root, and that too in addition to the vinculum. For an easy geometrical proof of the relation $x = 2z^2/(1+z^2)$, see Note 94.

[92] See Cantor, III, pp. 78-81. Also note the introduction of what is now a standard substitution in integration for the purpose of rationalization.

of universal application. Moreover it is evident that in this way
the arc may be obtained from the sine, and other things of
the same kind, but indirectly. So when later he heard that these
things had been derived in a direct manner by Newton with the
help of root-extractions,[93] he was desirous of getting a knowledge
of the matter.

From the above it was at once apparent that, using the method
by which Nicolaus Mercator had given the arithmetical tetragonism
of the hyperbola by means of an infinite series, that of the circle
might also be given, though not so symmetrically, by dividing by
$1 + zz$, in the same way that the former had divided by $1 + z$. The
author, however, soon found a general theorem for the area of any
central conic. Namely, the sector included by the arc of a conic
section, starting from the vertex, and two straight lines joining
its ends to the center, is equal to the rectangle contained by the
semi-transverse axis and a straight line of length

$$t \pm \frac{1}{3} t^3 + \frac{1}{5} t^5 \pm \frac{1}{7} t^7 + \ldots\ldots, \quad {}^{94}$$

where t is the portion of the tangent at the vertex intercepted
between the vertex and the tangent at the other extremity of the
arc, and unity is the square on the semi-conjugate axis or the rectangle contained by the halves of the latus-rectum and the transverse
axis, and $\pm$ is to be taken to mean $+$ for the hyperbola and $-$ for the
circle or the ellipse. Hence if the square of the diameter is taken to
be unity, then the area of the circle is

$$\frac{1}{1} - \frac{1}{3} + \frac{1}{5} - \frac{1}{7} + \frac{1}{9} - \frac{1}{11} + \text{etc.}$$

[93] This term represents what is now generally known as the method of
inversion of series. Thus, if we are given

$$x = y + ay^2 + by^3 + cy^4 + \text{etc.,}$$

where x and y are small, then $y = x$ is a first approximation; hence since
$y = x - ay^2 - by^3 - cy^4 - \text{etc.}$, we have as a second approximation

$$y = x - ax^2;$$

substituting this in the term containing y^2, and the first approximation, $y = x$,
in the term containing y^3, we have

$$y = x - a(x - ax^2)^2 - bx^3 = x - ax^2 + (2a^2 - b)x^3,$$

as a third approximation; and so on.

[94] The relation $x = 2z^2/(1 + z^2)$ can be easily proved geometrically for

When our friend showed this to Huygens, together with a proof of it, the latter praised it very highly, and when he returned the dissertation said, in the letter that accompanied it, that it would be a discovery always to be remembered among mathematicians, and that in it the hope was born that at some time it might be possible that the general solution should be obtained either by exhibiting its true value or by proving the impossibility of expressing it in recognized numbers.[95]   There is no doubt that neither he nor the discoverer, nor yet any one else in Paris, had heard anything at all by report concerning the expression of the area of a circle by means of an infinite series of rationals (such as afterward it became known had been worked out by Newton and Gregory).   Certainly Huygens did not, as is evident from the short

the circle; hence, by using the orthogonal projection theorem, Leibniz's result for the central conic can be immediately derived.

Thus suppose that, in the diagrams below, AC is taken to be unity, then $AU = z$ and $AX = x$.

Then, in either figure, since the $\triangle$s BYX, CUA are similar,

$$AX : XB = AX \cdot XB : XB^2 = XY^2 : XB^2 = AU^2 : CA^2 ;$$

hence, for the circle, we have

$$AX : AB = AU^2 : AC^2 + AU^2, \text{ or } x = 2z^2/(1 + z^2) ;$$

and similarly for the rectangular hyperbola

$$AX : AB = AU^2 : AC^2 - AU^2, \text{ or } x = 2z^2/(1 - z^2).$$

Applying all the x's to the tangent at A, we have (by division and integration of the right-hand side, term by term, in the same way as Mercator)

$$\text{area } AUMA = 2(z^3/3 \mp z^5/5 + z^7/7 \mp \text{ etc.})$$

Now, since the triangles UAC, YXB are similar, $UA \cdot XB = AC \cdot XY$; hence $2\triangle AYC = 2UA \cdot AC \mp UA \cdot AX = 2UA \cdot AC \mp AUMA \mp 2 \text{ seg. } AYA$, for Leibniz has shown that $AXMA = 2 \text{ seg. } AYA$; hence it follows immediately that

$$\text{sector } ACYA = z \mp z^3/3 + z^5/5 \mp \text{ etc.}$$

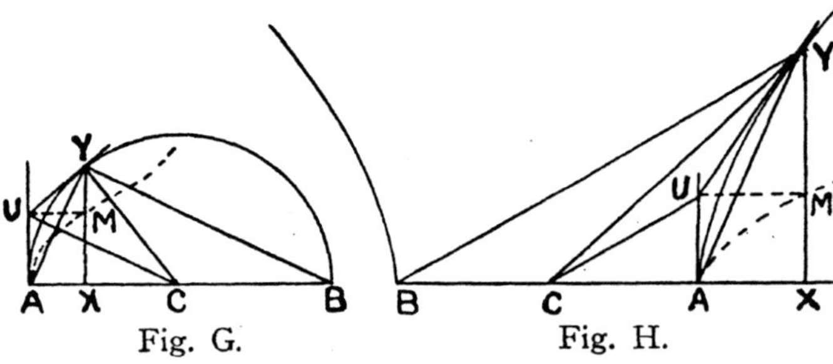

Fig. G.　　　　　Fig. H.

If now, keeping the vertical axis equal to unity, the transverse axis is made equal to a, Leibniz's general theorem follows at once from the orthogonal projection relation.

Note that z is, from the nature of the diagrams, less than 1.

[95] Wallis's expression for π as an infinite product, given in the *Arithmetica* (or Brouncker's derived expression in the form of an infinite continued fraction), or the argument used by Wallis in his work, could not possibly be taken as a proof that π could not be expressed in recognized numbers.

letter from him that I give herewith.[96]... Thus Huygens believed that it was now proved for the first time that the area of a circle was exactly equal to a series of rational quantities. Leibniz (relying on the opinion of Huygens, who was well versed in such matters), believed the same thing and so wrote those two letters to Oldenburg in 1674, which his opponents have published, in which he announces it as a new discovery;[97] indeed he went so far as to say that he, before all others, had found the magnitude of the circle expressed as a series of rational numbers, as had already been done in the case of the hyperbola.[98] Now, if Oldenburg had already communicated to him during his stay in London the series of Newton and Gregory,[99] it would have been the height of impudence for him to have dared to write in this way to Oldenburg; and either forgetfulness or collusion on the part of Oldenburg in not charging him with the deceit. For these opponents publish the reply of Oldenburg, in which he merely points out (he says "I do not wish you to be unaware....") that similar series had been noted by Gregory and Newton; and these things also he communicated in the year following in a letter (which they publish) written in the month of April.[100] From which it can be seen that they are blinded with envy or shameless with spite who dare to pretend that Oldenburg had already communicated those things to him in the preceding year. Yet there may be some blindness in their spite, because they do not see that they publish things by which their lying statements are refuted, nor that it would have been far better to have suppressed these letters between him and Oldenburg, as they have done in the case of others, either wholly or in part. Besides, from this time onwards he begins to correspond with Oldenburg about geometry; that is, from the time when he, who up till then had been but a

[96] The letter that is missing would no doubt have been given, in the event of the *Historia* being published. According to Gerhardt it is to be found in *Ch. Hugenii....exercitationes,* ed. Uylenbroeck, Vol. I, p. 6, under date Nov. 7, 1674.

[97] Collins wrote to Gregory in Dec. 1670, telling him of Newton's series for a sine, etc.; Gregory replied to Collins in Feb. 1671, giving him three series for the arc, tangent and secant; these were probably the outcome of his work on *Vera Circuli* (1667).

[98] By Mercator; query, also an allusion to Brouncker's article in the *Phil. Trans.,* 1668.

[99] Quite conclusive; no other argument seems required.

[100] This date, April 12, 1675, is important; it marks the time when Leibniz first began to speak of geometry in his correspondence with Oldenburg, as he says below.

beginner in this subject, first found out anything that he considered worthy to be communicated; and former letters written from Paris on March 30, April 26, May 24, and June 8, in the year 1673, which they say they have at hand but suppress, together with the replies of Oldenburg, must undoubtedly have dealt with other matters and have nothing in them to render those fictitious communications from Oldenburg the more deserving of belief. Again, when our young friend heard that Newton and Gregory had discovered their series by the extraction of roots,[101] he acknowledged that this was new to him, nor at first did he understand it very much; and he confessed as much quite frankly and asked for information on certain points, especially for the case in which reciprocal series were sought, by means of which from one infinite series the root was extracted by means of another infinite series. And from this also it is evident that what his opponents assert, that Oldenburg communicated the writings of Newton to him, is false; for if that were the truth, there would have been no need to ask for further information. On the other hand, when he began to develop his differential calculus, he was convinced that the new method was much more universal for finding infinite series without root-extractions, and adapted not only for ordinary quantities but for transcendent quantities as well, by assuming that the series required was given; and he used this method to complete his short essay on the arithmetical quadrature; in this he also included other series that he had discovered, such as an expression for the arc in terms of the sine or the complement of the sine, and conversely he showed how, by this same method, to find the sine or cosine when the arc was given.[102] This too is the reason why later he stood in no need of other methods than his own; and finally, he published his own new way of obtaining series in the *Acta Eruditorum.* Moreover, as it was at this time, just after he had published the essay on the Arithmetical Quadrature in Paris, that he was

---

[101] Newton obtained the series for arcsin x from the relation $\dot{a} : \dot{x} = 1 : \sqrt{(1 - x^2)}$, by expansion and integration, and then the series for the sine by the "extraction of roots." See Note 93, and, for Newton's own modification, Cantor, III, p. 73.

[102] It would appear from this that Leibniz could differentiate the trigonometrical functions. Professor Love, on the authority of Cantor, ascribes them to Cotes; but I have shown in an article in *The Monist* for April, 1916, that Barrow had explicitly differentiated the tangent and that his figures could be used for all the other ratios. Note the word "later" in the next sentence.

recalled to Germany, having perfected the technique of the new calculus he paid less attention to the former methods.

Now it is to be shown how, little by little, our friend arrived at the new kind of notation that he called the differential calculus. In the year 1672, while conversing with Huygens on the properties of numbers, the latter propounded to him this problem:[103] To find the sum of a decreasing series of fractions, of which the numerators are all unity and the denominators are the triangular numbers; of which he said that he had found the sum among the contributions of Hudde on the estimation of probability. Leibniz found the sum to be 2, which agreed with that given by Huygens. While doing this he found the sums of a number of arithmetical series of the same kind in which the numbers are any combinatory numbers whatever, and communicated the results to Oldenburg in February 1673, as his opponents have stated. When later he saw the Arithmetical Triangle of Pascal, he formed on the same plan his own Harmonic Triangle.[104]

Arithmetical Triangle

in which the fundamental series is an arithmetical progression

1, 2, 3, 4, 5, 6, 7, . . .

$$
\begin{array}{ccccccccc}
 & & & & 1 & & & & \\
 & & & 1 & & 1 & & & \\
 & & 1 & & 2 & & 1 & & \\
 & 1 & & 3 & & 3 & & 1 & \\
1 & & 4 & & 6 & & 4 & & 1 \\
\end{array}
$$

1   5   10   10   5   1

1   6   15   20   15   6   1

1   7   21   35   35   21   7   1

[103] Probably only to test Leibniz's knowledge.

[104] Gerhardt states that in the first draft of the *Historia*, Leibniz had bordered the Harmonic Triangle, as given here, with a set of fractions, each equal to 1/1, so as to correspond more exactly with the Arithmetical Triangle.

Harmonic Triangle

in which the fundamental series is a harmonical progression;

$$\frac{1}{1}$$

$$\frac{1}{2} \qquad \frac{1}{2}$$

$$\frac{1}{3} \qquad \frac{1}{6} \qquad \frac{1}{3}$$

$$\frac{1}{4} \qquad \frac{1}{12} \qquad \frac{1}{12} \qquad \frac{1}{4}$$

$$\frac{1}{5} \qquad \frac{1}{20} \qquad \frac{1}{30} \qquad \frac{1}{20} \qquad \frac{1}{5}$$

$$\frac{1}{6} \qquad \frac{1}{30} \qquad \frac{1}{60} \qquad \frac{1}{60} \qquad \frac{1}{30} \qquad \frac{1}{6}$$

$$\frac{1}{7} \qquad \frac{1}{42} \qquad \frac{1}{105} \qquad \frac{1}{140} \qquad \frac{1}{105} \qquad \frac{1}{42} \qquad \frac{1}{7}$$

where, if the denominators of any series descending obliquely to infinity or of any parallel finite series, are each divided by the term that corresponds in the first series, the combinatory numbers are produced, namely those that are contained in the arithmetical triangle. Moreover this property is common to either triangle, namely, that the oblique series are the sum- and difference-series of one another. In the Arithmetical Triangle any given series is the sum-series of the series that immediately precedes it, and the difference-series of the one that follows it; in the Harmonic Triangle, on the other hand, each series is the sum-series of the series following it, and the difference-series of the series that precedes it. From which it follows that

$$\frac{1}{1} + \frac{1}{2} + \frac{1}{3} + \frac{1}{4} + \frac{1}{5} + \frac{1}{6} + \frac{1}{7} + \text{etc.} = \frac{1}{0}$$

$$\frac{1}{1} + \frac{1}{3} + \frac{1}{6} + \frac{1}{10} + \frac{1}{15} + \frac{1}{21} + \frac{1}{28} + \text{etc.} = \frac{2}{1}$$

$$\frac{1}{1} + \frac{1}{4} + \frac{1}{10} + \frac{1}{20} + \frac{1}{35} + \frac{1}{56} + \frac{1}{84} + \text{etc.} = \frac{3}{2}$$

$$\frac{1}{1} + \frac{1}{5} + \frac{1}{15} + \frac{1}{35} + \frac{1}{70} + \frac{1}{126} + \frac{1}{210} + \text{etc.} = \frac{4}{3}$$

and so on.

Now he had found out these things before he had turned to Cartesian analysis; but when he had had his thoughts directed to this, he considered that any term of a series could in most cases be denoted by some general notation, by which it might be referred to some simple series. For instance, if the general term of the series of natural numbers is denoted by x, then the general term of the series of squares would be x^2, that of the cubes would be x^3, and so on. Any triangular number, such as 0, 1, 3, 6, 10, would be

$$\frac{x \cdot x + 1}{1 \cdot 2} \text{ or } \frac{xx + x}{2},$$

any pyramidal number, such as 0, 1, 4, 10, 20, etc., would be

$$\frac{x \cdot x + 1 \cdot x + 2}{1 \cdot 2 \cdot 3} \text{ or } \frac{x^3 + 3xx + 2x}{6},$$

and so on.

From this it was possible to obtain the difference-series of a given series, and in some cases its sum as well, when it was expressed numerically. For instance, the square is xx, the next greater square is $xx + 2x + 1$, and the difference of these is $2x + 1$; i. e., the series of odd numbers is the difference-series for the series of squares. For, if x is 0, 1, 2, 3, 4, etc., then $2x + 1$ is 1, 3, 5, 7, 9. In the same way the difference between x^3 and $x^3 + 3xx + 3x + 1$ is $3xx + 3x + 1$, and thus the latter is the general term of the difference-series for the series of cubes. Further, if the value of the general term can thus be expressed by means of a variable x so that the variable does not enter into a denominator or an exponent, he perceived that he could always find the sum-series of the given series. For instance, to find the sum of the squares, since it is plain that the variable cannot be raised to a higher degree than the cube, he supposed its general term z to be

$$z = lx^3 + mxx + nx, \text{ where } dz \text{ has to be } xx;$$

we have $dz = l\, d(x^3) + m\, d(xx) + n$, (where dx is taken $= 1$); now $d(x^3) = 3xx + 3x + 1$, and $d(xx) = 2x + 1$, as already found; hence

$$dz = 3lxx + 3lx + l + 2mx + m + n \backsimeq xx;^{105}$$

therefore $l = \dfrac{1}{3}$, $m = -\dfrac{1}{2}$, and $\dfrac{1}{3} - \dfrac{1}{2} + n = 0$, or $n = \dfrac{1}{6}$;

---

[105] The sign here used appears to be an invention of Leibniz to denote an identity, such as is denoted by $\equiv$ at present.

and the general term of the sum-series for the squares is

$$\frac{1}{3}x^3 - \frac{1}{2}xx + \frac{1}{6}x \text{ or } 2x^3 - 3xx + x, : 6. \text{ [106]}$$

As an example, if it is desired to find the sum of the first nine or ten squares, i. e., from 1 to 81 or from 1 to 100, take for x the values 10 or 11, the numbers next greater than the root of the last square, and $2x^2 - 3xx + x, : 6$ will be $2000 - 300 + 10, : 6 = 285$, or $2.1331 - 3.121 + 11, : 6 = 385$. Nor is it much more difficult with this formula to sum the first 100 or 1000 squares. The same method holds good for any powers of the natural numbers or for expressions which are made up from such powers, so that it is always possible to sum as many terms as we please of such series by a formula. But our friend saw that it was not always easy to proceed in the same way when the variable entered into the denominator, as it was not always possible to find the sum of a numerical series; however, on following up this same analytical method, he found in general, and published the result in the *Acta Eruditorum*, that a sum-series could always be found, or the matter be reduced to finding the sum of a number of fractional terms such as $1/x$, $1/xx$, $1/x^3$, etc, which at any rate, if the number of terms taken is finite, can be summed, though hardly in a short way (as by a formula); but if it is a question of an infinite number of terms, then terms such as $1/x$ cannot be summed at all, because the total of an infinite number of terms of such a series is an infinite quantity, but that of an infinite number of terms such as $1/xx$, $1/x^3$, etc., make a finite quantity, which nevertheless could not up till now be summed, except by taking quadratures. So, in the year 1682, in the month of February, he noted in the *Acta Eruditorum* that if the numbers $1.3, 3.5, 5.7, 7.9, 9.11$, etc., or $3, 15, 35, 63, 99$, etc., are taken, and from them is formed the series of fractions

$$\frac{1}{3} + \frac{1}{15} + \frac{1}{35} + \frac{1}{63} + \frac{1}{99} + \text{ etc.,}$$

then the sum of this series continued to infinity is nothing else but $\frac{1}{2}$; while, if every other fraction is left out, $\frac{1}{3} + \frac{1}{35} + \frac{1}{99} + \text{etc.}$

[106] This, and other formulas of the same kind, had been given by Wallis in connection with the formulas for the sums of the figurate numbers. Wallis called these latter sums the "characters" of the series.

expresses the magnitude of a semicircle of which the square on the diameter is represented by 1.[107]

Thus, suppose $x = 1, 2, 3$, etc.[108] Then the general term of

$$\frac{1}{3} + \frac{1}{15} + \frac{1}{35} + \frac{1}{63} + \text{etc. is } \frac{1}{4xx + 8x + 3};$$

it is required to find the general term of the sum-series.

Let us try whether it can have the form $e/(bx + c)$, the reasoning being very simple; then we shall have

$$\frac{e}{bx + c} - \frac{e}{bx + b + c} = \frac{eb}{bbxx + bbx + bc + 2bcx + cc} \backsim \frac{1}{4xx + 8x + 3};$$

hence, equating coefficients in these two formulas, we have

$$b = 2,\ eb = 1,\ \text{or } e = \tfrac{1}{2},$$
$$bb + 2bc = 8,\ \text{or } 4 + 4c = 8,\ \text{or } c = 1;$$

and finally we should have also $bc + cc = 3$, which is the case. Hence the general term of the sum-series is $(1:2)/(2x + 1)$ or $1/(4x + 2)$, and these numbers of the form $4x + 2$ are the doubles of the odd numbers. Finally he gave a method for applying the differential calculus to numerical series when the variable entered into the exponent, as in a geometrical progression, where, taking any radix b the term is b^x, where x stands for a natural number. The terms of the differential series will be $b^{x+1} - b^x$, or $b^x(b-1)$; and from this it is plain that the differential series of the given geometrical series is also a geometrical series proportional to the given series. Thus the sum of a geometrical series may be obtained.

But our young friend quickly observed that the differential calculus could be employed with diagrams in an even more wonderfully simple manner than it was with numbers, because with diagrams the differences were not comparable with the things which differed; and as often as they were connected together by addition or subtraction, being incomparable with one another, the less vanished in comparison with the greater; and thus irrationals could be differentiated no less easily than surds, and also, by the aid of logarithms, so could exponentials. Moreover, he observed that the infinitely small lines occurring in diagrams were nothing else but the

[107] This sentence, in that it breaks the sense from the preceding sentence to the one that follows, would appear to be an interpolated note.

[108] There is an unimportant error here. The first value of x evidently should be 0, and not 1.

momentaneous differences of the variable lines. Also, in the same way as quantities hitherto considered by analytical mathematicians had their functions such as powers and roots, so also such quantities as were variable had new functions, namely, differences. Also, that as hitherto we had x, xx, x^3, etc., y, yy, y^3, etc., so now it was possible to have dx, ddx, d^3x, etc., dy, ddy, d^3y, and so forth. In the same way, that it was possible to express curves, which Descartes had excluded as being "mechanical," by equations of position, and to apply the calculus to them and thus to free the mind from a perpetual reference to diagrams. In the applications of the differential calculus to geometry, differentiations of the first degree were equivalent to nothing else but the finding of tangents, differentiations of the second degree to the finding of osculating circles (the use of which was introduced by our friend) ; and that it was possible to go on in the same fashion. Nor were these things only of service for tangents and quadratures, but for all kinds of problems and theorems in whch the differences were intermingled with integral terms (as that brilliant mathematician Bernoulli called them), such as are used in physico-mechanical problems.

Thus it follows generally that if any series of numbers or lines of a figure have a property that depends on two, three or more consecutive terms, it can be expressed by an equation involving differences of the first, second, third, or higher degree. Moreover, he discovered general theorems for any degree of the differences, just as we have had theorems of any degree, and he made out the remarkable analogy between powers and differences published in the *Miscellania Berolinensia*.

If his rival had known of these matters, he would not have used dots to denote the degrees of the differences,[109] which are useless for expressing the general degree of the differences, but would have used the symbol d given by our friend or something similar, for then d^e can express the degree of the difference in general. Besides everything which was once referred to figures, can now be expressed by the calculus.

---

[109] Why not? Newton's dotted letters still form the best notation for a certain type of problem, those which involve equations of motion in which the independent variable is the time, such as central orbits. Probably Leibniz would class the suffix notation as a variation of his own, but the D-operator eclipses them all. For beginners, whether scholastic or historically such (like the mathematicians that Barrow, Leibniz and Newton were endeavoring to teach), the separate letter notation has most to recommend it on the score of ease of comprehension; we find it even now used in partial differential equations.

For $\sqrt{(dxdx+dydy)}$[110] is the element of the arc of a curve, ydx is the element of its area; and from that it is immediately evident that $\int y\,dx$ and $\int x\,dy$ are the complements of one another, since $d(xy) = x\,dy + y\,dx$, or conversely, $xy = \int x\,dy + \int y\,dx$, however these figures vary from time to time; and from this, since $xyz = \int xy\,dz + \int xz\,dy + \int yz\,dx$, three solids are also given that are complementary, every two to the third. Nor is there any need for him to have known those theorems which we deduced above from the characteristic triangle; for example, the moment of a curve about the axis is sufficiently expressed by $\int x\sqrt{(dxdx+dydy)}$. Also what Gregory St. Vincent has concerning *ductus*, what he or

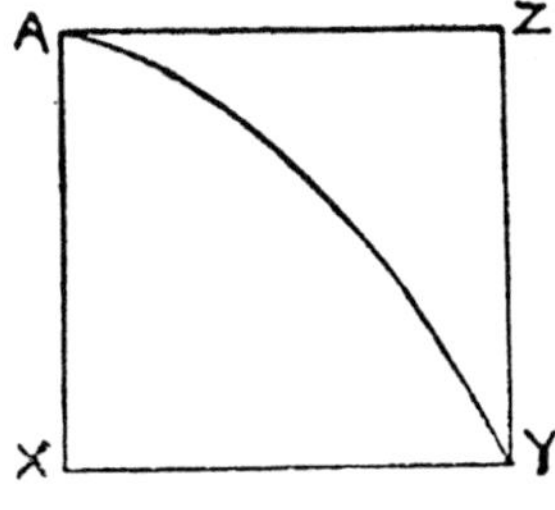

Fig. 5.

Pascal had concerning *ungulae* and *cunei*,[111] every one of these is immediately deduced from a calculus such as this. Thus Leibniz saw with delight those discoveries that he had applauded in others obtained by himself, and thereupon he left off studying them at all closely, because all of them were contained in a calculus such as his.

For example, the moment of the figure AXYA (Fig. 5) about the axis is $\frac{1}{2}\int yy\,dx$, the moment of the figure about the tangent at the vertex is $\int xy\,dx$, the moment of the complementary trilinear figure AZYA about the tangent at the vertex is $\frac{1}{2}\int xx\,dy$. Now these two last moments taken together yield the moment of the circumscribed rectangle AXYZ about the tangent at the vertex, and are complementary to one another.

However, the calculus also shows this without reference to any figure, for $\frac{1}{2}d(xxy) = xy\,dx + \frac{1}{2}xx\,dy$; so that now there is need

---

[110] Leibniz does not give us an opportunity of seeing how he would have written the equivalent of $dxdxdx$; whether as dx^3 or $\overline{dx}^3$ or $(dx)^3$.

[111] *Ductus* and *ungulae* have already been explained in Notes 28, 29; *cuneus* denotes a wedge-shaped solid; cf. "cuneiform."

for no greater number of the fine theorems of celebrated men for Archimedean geometry, than at most those given by Euclid in his Book II or elsewhere, for ordinary geometry.

It was good to find that thereafter the calculus of transcendent quantities should reduce to ordinary quantities, and Huygens[112] was especially pleased with this. Thus, if it is found that

$$2 \int \frac{dy}{y} = 3 \int \frac{dx}{x},$$

then from this we get $yy = x^3$, and this too from the nature of logarithms combined with the differential calculus, the former also being derived from the same calculus. For let $x^m = y$, then $mx^{m-1}\,dx = dy$. Hence, dividing each side by equal things, we have

$$m\,\frac{dx}{x} = \frac{dy}{y}.$$

Again, from the equation, $m \log x = \log y$, we have

$$\log x : \log y = \int \frac{dx}{x} : \int \frac{dy}{y}.^{113}$$

By this the exponential calculus is rendered practicable as well. For let $y^x = z$, then $x \log y = \log z$, $dx \log y + x\,dy : y = dz : z$.

In this way we free the exponents from the variable, or at other times we may transpose the variable exponent with advantage under the circumstances. Lastly, those things that were once held in high esteem are thus made a mere child's-play.

Now of all this calculus not the slightest trace existed in all the writings of his rival before the principles of the calculus were

[112] This is peculiar. The demonstration that follows was beyond the powers of Leibniz in June, 1676 (see pp. 121, 122), probably so until Nov., 1676, when he was in Holland, and possibly later still. Hence the result would have been communicated to Huygens by letter, and there would be an answer from Huygens. I have been so far unable to find such a letter.

[113] This only proves the proportionality, enabling Leibniz to convert the equation $2\int dy/y = 3\int dx/x$ into $2 \log y = 3\log x$. It will hardly suffice as it stands to enable him to deal with such an equation as $2\int dy/y = 3\int x\,dx$; and it is to be noted that Leibniz does not notice at all the constant of integration. Although Barrow has in effect differentiated (and therefore also has the inverse integral theorems corresponding thereto) both a logarithm and an exponential in Lecture XII, App. III, Prob. 3, 4, yet these problems are in such an ambiguous form that it may be doubted whether Barrow was himself quite clear on what he had obtained. Hence this clear statement of Leibniz must be considered as a great advance on Barrow.

published by our friend;[114] nor indeed anything at all that Huygens or Barrow had not accomplished in the same way, in the cases where they dealt with the same problems.

But how great was the extent of the assistance afforded by the use of this calculus was candidly acknowledged by Huygens; and this his opponents suppress as much as ever they can, and straightway go on with other matters, not mentioning the real differential calculus in the whole of their report. Instead, they adhere to a large extent to infinite series, the method for which no one denies that his rival brought out in advance of all others. For those things which he said enigmatically, and explained at a much later date, are all they talk about, namely, fluxions and fluents, i. e., finite quantities and their infinitely small elements; but as to how one can be derived from the other they offer not the slightest suggestion. Moreover, while he considers nascent or evanescent ratios, leading straight away from the differential calculus to the method of exhaustions, which is widely different from it (although it certainly also has its own uses), he proceeds not by means of the infinitely small, but by ordinary quantities, though these latter do finally become the former.

Since therefore his opponents, neither from the *Commercium Epistolicum* that they have published, nor from any other source, brought forward the slightest bit of evidence whereby it might be established that his rival used the differential calculus before it was published by our friend; therefore all the accusations that were brought against him by these persons may be treated with contempt as beside the question. They have used the dodge of the pettifogging advocate[115] to divert the attention of the judges from the matter on trial to other things, namely to infinite series. But even in these they could bring forward nothing that could impugn the honesty of our friend, for he plainly acknowledged the manner in which he had made progress in them; and in truth in these also, he finally attained to something higher and more general.

[114] Almost seems to read as a counter-charge against Newton of stealing Leibniz's calculus. Note the tardy acknowledgement that Barrow has previously done all that Newton had given.

[115] The whole effect that this *Historia* produces in my mind is that the entire thing is calculated to the same end as the *Commercium Epistolicum.* The pity of it is that Leibniz could have told such a straightforward tale, if events had been related in strict *chronological* order, without any interpolations of results that were derived, or notation that was perfected, later. A tale so told would have proved once and for all how baseless were the accusations of the *Commercium,* and largely explained his denial of any obligations to Barrow.

SUPPLEMENT.

Barrow, *Lectiones Geometricae*, Lect. XII, Prop. 1, 2, 3.

[Page 105, First Edition, 1670.]

General foreword. We will now proceed with the matter in hand; and, in order that we may save time and words, it is to be observed everywhere in what now follows that AB is some curved line, such as we shall draw, of which the axis is AD; to this axis all the straight lines BD, CA, MF, NG are applied perpendicular; the arc MN is indefinitely small; the straight line $\alpha\beta =$ arc AB, the straight line $\alpha\mu =$ arc AM, and $\mu\nu =$ arc MN; also lines applied to $\alpha\beta$ are perpendicular to it. On this understanding:

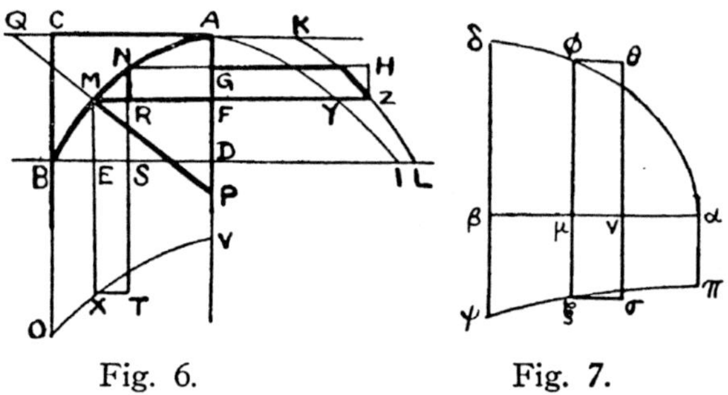

Fig. 6.   Fig. 7.

1. Let MP be perpendicular to the curve AB, and the lines KZL, $\alpha\phi\delta$ such that FZ $=$ MP, $\mu\phi =$ MF. Then the spaces $\alpha\beta\delta$, ADLK are equal.
For the triangles MRN, PFM *are similar,* MN : NR $=$ PM : MF,

$$MN.MF = NR.PM;$$

that is, on substituting the equal quantities,

$$\mu\nu.\mu\phi = FG.FZ, \text{ or rect. } \mu\theta = \text{ rect. FH.}$$

But the space $\alpha\beta\delta$ only differs in the slightest degree from an infinite number of rectangles such as $\mu\theta$, and the space ADLK is equivalent to an equal number of rectangles such as FH. Hence the proposition follows.

2. Hence, *if the curve AMB is rotated about the axis* AD, *the ratio of the surface produced to the space* ADLK *is that of the circumference of a circle to its diameter*; whence if the space ADLK is known, the said surface is known.
Some time ago I assigned the reason why this was so.

3. Hence, *the surfaces of the sphere, both the spheroids, and the conoids receive measurement.* For if AD is the axis of the conic section, etc.

NOTE. In the above figure, I have "lined in" the part of the diagram (which serves for about ten theorems) especially used for the first two theorems. If this is compared with Leibniz's figure on p. 15, further comment is needless.

IV.

MANUSCRIPTS OF THE PERIOD 1673-1675.

§ 3.*

The following notes, on certain MSS. which Gerhardt does not give in full, are taken from G. 1848, p. 20 et seq. (see also G. 1855, pp. 55 et seq.).

In a manuscript of August, 1673, bearing the title *Methodus nova investigandi Tangentes linearum curvarum ex datis applicatis, vel contra Applicatis ex datis productis, reductis, tangentibus, perpendicularibus, secantibus*, Leibniz begins at once with an attempt to find a method that is applicable to any curve for the determination of its tangent. "But if," says Leibniz with regard to the classification of curves which Descartes laid down as fundamental for his method of tangents, "the figure is not geometrical — such as the cycloid—it does not matter; for it will be treated as an example of a geometrical curve, by supposing that there is a relation between the straight lines and curves by which they are made known to us; in this way, tangents can be drawn just as well to either geometrical or ageometrical curves, as far as the nature of the figure allows." He considers the curve as a polygon with an infinite number of sides, and here already he constructs what he calls the "Characteristic Triangle," whose sides are an infinitely small arc of the curve, and the differences between the ordinates and between the abscissae; this is similar to the triangle whose sides are the tangent, the subtangent and the ordinate for the point of contact. In just the same manner as used by Descartes, Leibniz seeks the tangent by means of the subtangent; he denotes the infinitely small differences of the abscissae by b, and verifies for the parabola, that his method works out correctly, when the terms of the equation that contain the in-

* §§ 3-10 inclusive appeared in *The Monist* for April, 1917.

finitely small quantities are neglected. The omission of these terms, however, does not appear to Leibniz to be a method to be relied upon. In fact, he says: "It is not safe to reject multiples of the infinitely small part b, and other things; for it may happen that through the compensation of these with others,[1] the equation may come to a totally different condition." So he seeks to obtain the determination of the subtangent in some other way. "The whole question is, how the applied lines can be found from the differences of two applied lines," are his own words. He then finds that the solution of this problem reduces to the summation of a series, of which the terms are the differences of consecutive abscissae.

At the end of the manuscript Leibniz proceeds to speak of the inverse problem: "It is an important subject for investigation, whether it is possible, by retracing our steps, to proceed from tangents and other functions to ordinates. The matter will be most accurately investigated by tables[2] of equations; in this way we may find out in how many ways some one equation may be produced from others, and from that, which of them should be chosen in any case. This is, as it were, an analysis of the analysis itself, but if that is done it forms the fundamental of human science, as far as this kind of things is concerned." Ultimately Leibniz obtains the following result: "The two questions, the first that of finding the description of the curve from its elements, the second that of finding the figure from the given differences, both reduce to the same thing. From this fact it can be taken that almost the whole of the theory of the inverse method of tangents is reducible to quadratures."

According to this, Leibniz has in the middle of the year 1673 already attained to the knowledge that the direct and the so-called inverse tangent-problem have an undoubted connection with one another; he has an idea that the latter may be capable of reduction to a quadrature (i. e., to a summation).

Again, in a manuscript dated October 1674, i. e., fourteen months later, which bears the title *Schediasma de Methodo Tangentium inversa ad circulum applicata,* he is able to say for certain that "the quadratures of all figures follow from the inverse method

---

[1] It is impossible to see, without a fuller knowledge of the context, whether this refers to "compensation of errors," or whether Leibniz is alluding to the possibility of all the finite terms cancelling one another.

[2] Leibniz comes back to this point later; see § 5.

of tangents, and thus the whole science of sums and quadratures can be reduced to analysis, a thing that nobody even had any hopes of before."

After Leibniz thus recognized the identity between the inverse tangent-problem, of which the general solution had not been found by Descartes, and the quadrature of curves, he applied himself to the investigation of series, by the summation of which quadratures were then obtained. In a very extensive discussion, bearing the date of October, 1674, and the title *Schediasma de serierum summis, et seriebus quadraticibus,* Leibniz starts from the series

$$\frac{b}{1} - \frac{b^2}{2} + \frac{b^3}{3} - \frac{b^4}{4} + \frac{b^5}{5} - \cdots,$$

and obtains the following general rule: "By calling the variable ordinates x, and the variable abscissae y, and b the abscissa of the greatest ordinate e, and d the abscissa of the least ordinate h," are Leibniz's own words, "we have the following rules:

$$\frac{x^2}{2} = ywx - \frac{yw^2}{2} + \frac{d^2h}{2},$$

$$\frac{h^2 w}{2} + \frac{d^2 h}{2} = xy - \frac{x}{2}, \quad e - h = w,$$

$$xw = \frac{e^2}{2} - \frac{w^2}{2},$$

$yw = x$ in decreasing values, for in ascending or increasing values $yw = eb - x$."[3]

Leibniz then goes on to remark: "These rules are to be altered slightly according as the series increase or decrease; also mention of the least ordinate may be omitted, if it is always understood to be the last ordinate; on the other hand, w can always be inserted wherever mention is made of w. All series hitherto found are contained in the one by means of these rules, except the series of powers, which is to be obtained by taking differences."

---

[3] This, without either proof or figure, is a hopeless muddle; and yet it is repeated word for word, without any addition or remark, in Gerhardt's 1855 publication. Goodness knows what the use of it was supposed to be in this form! Unless Leibniz has omitted some length, which he has supposed to be unity, the dimensions are all wrong.

In the same essay, Leibniz makes use of a theorem, which he has probably found to be general at an earlier date, namely:

"Since BC is to BD as WL to SW, therefore BC⌢SW,[4] that is, the sum of every BC [applied to AC], is equal to BD⌢WL, that is, the sum of every BD applied to the base; moreover, the sum of every BD applied to the base is equal to half the square on the greatest BD. Further, it is evident that the sum of every WL is equal to the greatest BD."

Accordingly, Leibniz comes to the further conclusion that the method of Descartes, which uses a subsidiary equation with two equal roots, to solve the general inverse-tangent problem, is unsatisfactory. In a manuscript of January, 1675, Leibniz says: "Thus at last I am free from the unprofitable hope of finding sums of series and quadratures of figures by means of a pair of equal roots, and I have discovered the reason why this argument cannot be used; this has worried me for quite long enough."[5]

§ 4.

The manuscript that comes next in date is one that is given in G. 1855. It really consists of three short notes, (1) a theorem on moments, (2) a continuation of the idea started at the end of the manuscript of August, 1673 (§ 3), namely the formation of tables of equations that are derivable from certain standard equations, with the appropriate substitutions for each case, (3) a return to the consideration of moments.

This is the first appearance of the word "moment," but from the context it is evident that Leibniz has done some considerable amount of work upon the idea before. If the theorem that is first given is written in modern notation,

---

[4] The sign ⌢ signifies multiplication.

[5] Observe that as yet nothing has been said about the area of surfaces of revolution or moments about the axis, although we should expect them to be mentioned in connection with the figure that is given; for the next manuscript shows that in October 1675, Leibniz has already done a considerable amount of work on moments.

it takes the form of an "integration by parts" and serves to change the independent variable. Thus we have

$$\int xy \, dx = \left[\frac{x^2y}{2}\right] - \int \frac{x^2}{2} \, dy;$$

and it is readily seen that if x can be expressed as a square root of a simple function of y, as for the circle and the conic sections, then the integral on the right-hand side has no irrationality. This, I take it, is the connection between this theorem and those which follow.

The proof is not so clear as it might be on account of two errors, both I think errors of transcription or misprints. The first a should be an x, and the second a should be the preposition a ($=$ from); also, for modern readers the figure might be improved by showing the *variable* lines AB ($= x$), BC ($= y$) as in the accompanying diagram. The argument then is as follows:

Moment of BC($=y$) about AD is xy, when it is applied to AB for the summation; for this brings in the infinitesimal breadth of the line.

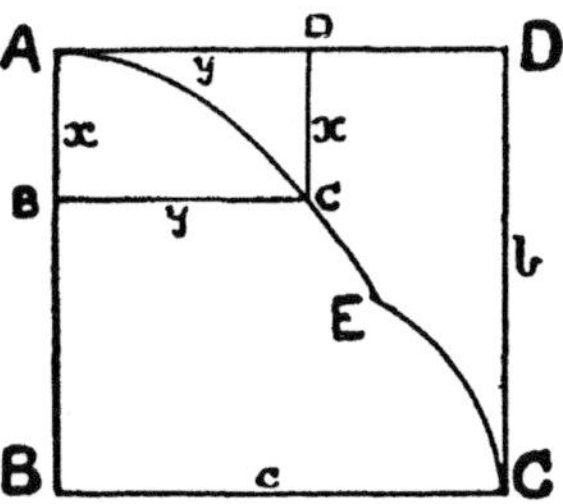

Moment of DC ($=x$) about AD is $x^2/2$, when applied to AD, so as to include the infinitesimal breadth of the line, and assuming that the line may be considered to be condensed at its center of gravity. The theorem follows at once.

Note the use of the sign $\sqcap$ as a symbol of equality, which I have allowed to stand in the opening paragraph. Leibniz adopts the ordinary sign two months later, or Gerhardt makes the change,[6] so I have not thought it necessary to adhere to it, but only to show it in the opening paragraph.

The only remark that seems to be necessary with regard to the second part of this manuscript is that Weissenborn[7] argues from the continued allusion by Leibniz to the desirability of forming tables of curves whose quadratures may be derived from those of others, especially the conic sections, (starting with the manuscript of November, 1675, where Weissenborn states that it is first hinted), that Leibniz had probably either seen or heard of the *Catalogus curvarum ad conicas sectiones relatarum* of Newton. The point is that Weissenborn seems to have missed the clear reference to the reduction of curves to those of the second degree, in this manuscript of October, 1675. It may of course be just possible that G. 1855, in which this MS. appears, was not at Weissenborn's hand at the time that he wrote, for Weissenborn's book was published in 1856.

With regard to the third part, it will be found in the original Latin that Leibniz, after apparently starting with perfect clearness, gets rather into a muddle toward the end. This is however only apparent, being partly due to an inaccurate figure, and partly to what I am convinced is an error of transcription. This incorrect sentence makes Leibniz write apparently absolute nonsense; but if a correction is made according to the suggestion in the footnote, and reference is made to the corrected diagram that I have added on the right of the figure of Leibniz, as given by Gerhardt, then the proof given by Leibniz reads perfectly smoothly and sensibly.

[6] Gerhardt has a footnote to the effect that, as nearly as possible he has retained the exact form of this and the manuscripts that immediately follow; except in the matter of this one sign I have adhered to the form given by Leibniz.

[7] Weissenborn, *Principien der höheren Analysis,* Halle, 1856.

25 October, 1675.

Analysis Tetragonistica Ex Centrobarycis.

[Analytical quadrature by means of centers of gravity.]

Let any curve AEC be referred to a right angle BAD; let AB ⊓ DC ⊓ a,[8] and let the last x ⊓ b; also let BC ⊓ AD ⊓ y, and the last y ⊓ c. Then it is plain that

$$\text{omn. } \overline{yx} \text{ to } x = \frac{b^2 c}{2} - \text{omn. } \overline{\frac{x^2}{2}} \text{ to } y. \cdots\cdots (1)$$

For, the moment of the space ABCEA about AD is made up of rectangles contained by BC ($= y$) and AB ($= x$); also the moment about AD of the space ADCEA, the complement of the former is made up of the sum of the squares on DC halved $\left(= \frac{x^2}{2}\right)$; and if this moment is taken away from the whole moment of the rectangle ABCD about AD, i. e., from c into omn. x,[9] or from $\frac{b^2 c}{2}$, there 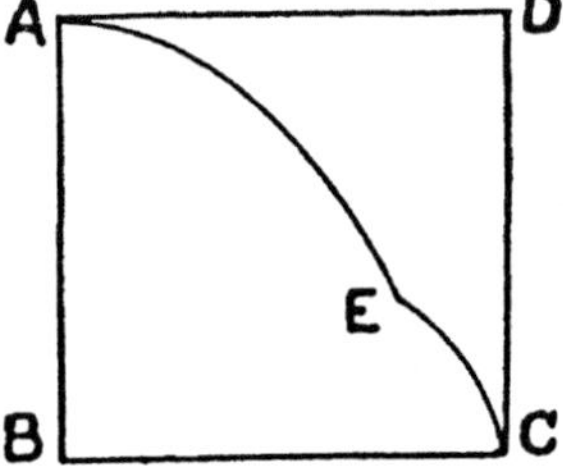 will remain the moment of the space ABCEA. Hence the equation that I gave is obtained; and, by rearranging it, it follows that

$$\text{omn. } yx \text{ to } x + \text{omn. } \frac{x^2}{2} \text{ to } y = \frac{b^2 c}{2} \cdots\cdots (2)$$

In this way we obtain the quadrature of the two joined in one in every case; and this is the fundamental theorem in the center of gravity method.

Let the equation expressing the nature of the curve be

$$ay^2 + bx^2 + cxy + dx + ey + f = 0, \cdots\cdots\cdots (3)$$

and suppose that $xy = z, \cdots (4)$, then $y = \frac{z}{x}. \cdots\cdots\cdots (5)$

Substituting this value in equation (3), we have

$$\frac{az^2}{x^2} + bx^2 + cz + dx + \frac{ez}{x} + f = 0, \cdots\cdots\cdots (6)$$

---

[8] This a should be x.

[9] Here, in the Latin, "*ac* in omn.x" should be "a c in omn.x."

and, on removing the fractions,

$$az^2 + bx^4 + cx^2z + dx^3 + exz + fx^2 = 0. \qquad (7)$$

Again, let $x^2 = 2w$ .... (8); then, substituting this value in equation (3), we have

$$ay^2 + 2bw + cxy + dx + ey + f = 0, \qquad (9)$$

and therefore

$$x = \frac{-ay^2 - 2bw - ey - f}{cy + d}, \qquad (10)$$

$$= \sqrt{2w}; \qquad (11)$$

and, squaring each side, we have[10]

$$a^2y^2 + 4aby^2w + 2aey^3 + 2afy^2 + 4b^2w^2 + 4bewy + 4bfw$$
$$+ e^2y^2 + 2fey + f^2 - 2c^2y^2w - 4cdyw - 2d^2w = 0. \quad ..(12)$$

Now, if a curve is described according to equation (7), and also another according to equation (12), I say that the quadrature of the figure of the one will depend on the quadrature of the figure of the other, and *vice versa*.

If, however, in place of equation (3), we took another of higher degree, the third say, we should again have two equations in place of (7) and (12); and continuing in this manner, there is no doubt that a certain definite progression of equations (7) and (12) would be obtained, so that without calculation it could be continued to infinity without much trouble. Moreover, from one given equation to any curve, all others can be expressed by a general form, and from these the most convenient can be selected.

If we are given the moment of any figure about any two straight lines, and also the area of the figure, then we have its center of gravity. Also, given the center of gravity of any figure (or line) and its magnitude, then we have its moment about any line whatever. So also, given the magnitude of a figure, and its moments about any two given straight lines, we have its moment about any straight line. Hence also we can get many quadratures from a few given ones. Moreover, the moment of any figure about any straight line can be expressed by a general calculation.

The moment divided by the magnitude gives the distance of the center of gravity from the axis of libration.

[10] In view of this accurate bit of algebra, the faulty work in subsequent manuscripts seems very unaccountable.

Suppose then that there are two straight lines in a plane, given in position, and let them either be parallel or meet, when produced, in F. Suppose that the moment about BC is found to be equal to ba^2, and the moment about DE is found to be ca^2. Call the area of the figure v; then the distance of the center of gravity from the straight line BC, namely CG, is equal to $\dfrac{ba^2}{v}$, and its distance from the straight line DE, namely EH, is equal to $\dfrac{ca^2}{v}$; therefore CG is to EH as b is to c, or they are in a given ratio.[11]

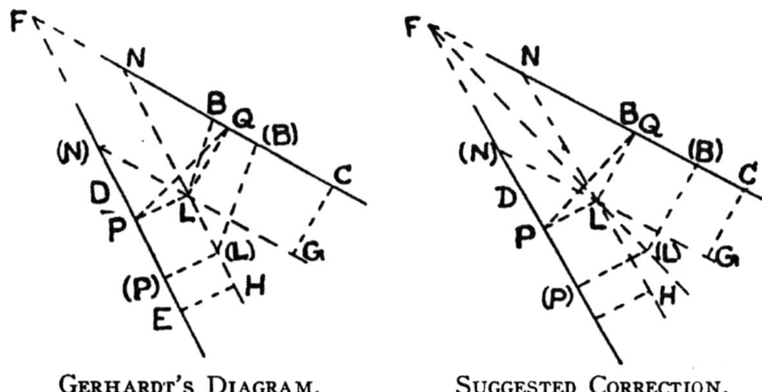

GERHARDT'S DIAGRAM.   SUGGESTED CORRECTION.

Now suppose that the straight line EH, remaining in the plane, traverses the straight line DE, always being perpendicular to it, and that the straight line CG traverses the straight line BC, always perpendicular to it, and that the end G leaves as it were its trace, the straight line G(N), and the end H the straight HN. Then, if BC and DE meet anywhere, G(N) and HN must also meet somewhere, either within or without the angle at F. Let them meet at L; then the angle HLG is equal to the angle EFC, and PLQ (supposing that PL = EH and LQ = CG) will be the supplement of the angle EFC between the two straight lines, and will thus be a given angle. If then PQ is joined, the triangle PQL is obtained, having a given vertical angle, and the ratio of the sides forming the vertex, QL : LP, also given.

When then BL is taken, or (B)(L), of any length whatever, since the angle BLP always remains the same, and in addition we have BL to LP as (B)(L) to (L)(P), therefore also BL to (B)(L) as LP to (L)(P) ; and this plainly happens when FL is also propor-

----

[11] This proves the fundamental theorem given lower down, with regard to a pair of parallel straight lines; and he now goes on to discuss the case of non-parallel straight lines.

tional to these, that is, when a straight line passes through F, L, (L),......

Hence, since we are not here given several regions, it follows that the locus is a straight line. Therefore, given the two moments of a figure about two straight lines that are not parallel,........., the area of the figure will be given, and also its center of gravity.[12]

Behold then the fundamental theorem on centers of gravity. If two moments of the same figure about two parallel straight lines are given, then the area of the figure is given, but not its center of gravity.

Since it is the aim of the center of gravity method to find dimensions from given moments, we have hence two general theorems:

If we are given two moments of the same figure about two straight lines, or axes of libration, that are parallel to one another, then its magnitude is given; also when the moments about three non-parallel straight lines are given. From this it is seen that a method for finding elliptic and hyperbolic curves from given quadratures of the circle and the hyperbola is evident.[13] But of this in a special note.

§ 5.

The next manuscript to be considered is a continuation of the preceding, and is dated the next day. Its character is of the nature of disjointed notes, set down for further consideration.

[12] The passage in Gerhardt reads:
Datis ergo duobus momentis figurae ex duabus rectis non parallelis, dabitur figurae momentis tribus axibus librationis, qui non sint omnes paralleli inter se, dabitur figurae area, et centrum gravitatis.
For this I suggest:
Datis ergo *tribus* momentis figurae ex *tribus* rectis non parallelis, *aliter* figurae momentis tribus axibus librationis, qui non *sunt* omnes paralleli inter se....
The passage would then read:
Given three moments of a figure about three straight lines that are not parallel, in other words, the moments of the figure about three axes of libration, which are not all parallel to one another, then the area of the figure will be given and also the center of gravity.
If the alternative words are *written* down, one under the other, and not too carefully, I think the suggested corrections will appear to be reasonable.

[13] Apparently, here Leibniz is referring back to the theorem at the beginning of the section.

26 October, 1675.

Another tetragonistic analysis can be obtained by the aid of curves. Thus, let the same curve be resolved into different elements, according as the ordinates are referred to different straight lines. Hence also arise diverse plane figures, consisting of elements similar to the given curve; and since all of these are to be found from the given dimension of the curve, it follows that from the dimension of any one of the curves of this kind the rest are obtained.

In other ways it is possible to obtain curves that depend on others, if to the given curve are added the ordinates of figures of which the quadrature is either known or can be obtained from the quadrature of the given one.

Just as areas are more easily dealt with than curves, because they can be cut up and resolved in more ways, so solids are more manageable than planes and surfaces in general. Therefore, whenever we divert the method for investigating surfaces to the consideration of solids, we discover many new properties; and often we may give demonstrations for surfaces by means of solids when they are with difficulty obtained from the surfaces themselves. Tschirnhaus observed in a delightful manner that most of the proofs given by Archimedes, such as the quadrature of the parabola, and dependent theorems on the sphere, cone, and cylinder, can be reduced to sections of rectilinear solids only, and to a composition that is easily seen and readily handled.

Various ways of describing new solids.

If from a point above a plane a rigid descending straight line is moved round an area, of any shape whatever, diverse kinds of conical bodies are produced. Thus if the plane area is bounded by the circumference of a circle, a right or scalene cone is produced. Also if the figure used for the base, or the plane area, has a center— an ellipse for example—then we get an elliptic cone, which is a right cone if the given point is directly above the center, and if not it is scalene. Another conic gives another elliptic cone.

If the rigid line drawn down from the point is circular or some other curve, at one time it is so fixed to the point or pole that it has freedom to move in one way only, say round an axis, in which case it is necessary that the base should be a circle and that the fixed point or pole should be directly over the center. At another time it is necessary that the rigid line should have freedom for other motions, such as an up and down motion, or some other motion, controlled by some straight line; and then it will always ascend or descend when necessary, so that it ever touches the given plane area by its rotation round the axis; and this is the second class of cones. A third class consists of those in which, besides the double motion of a rotation round an axis and an up and down motion, the curve alone, or the axis alone, or even both the curve and the axis, also perform other motions meanwhile, or even the point itself moves.

Here is another consideration.

The moments of the differences about a straight line perpendicular to the axis are equal to the complement of the sum of the terms; and the moments of the terms are equal to the complement of the sum of the sums, i. e.,

$$\text{omn.}\overline{xw} \sqcap \text{ult.}x,\ \overline{\text{omn.}w,,} - \overline{\text{omn. omn.}w} \qquad (14)$$

Let $xw \sqcap az$, then $w \sqcap \dfrac{az}{x}$, and we have

$$\text{omn.}az \sqcap \text{ult.}x,\ \text{omn.}\frac{az}{x} - \text{omn.omn.}\frac{az}{x}\ ;$$

hence

$$\text{omn.}\frac{az}{x} \sqcap \text{ult.}x\ \text{omn.}\frac{az}{x^2} - \text{omn.omn.}\frac{az}{x^2}\ ;$$

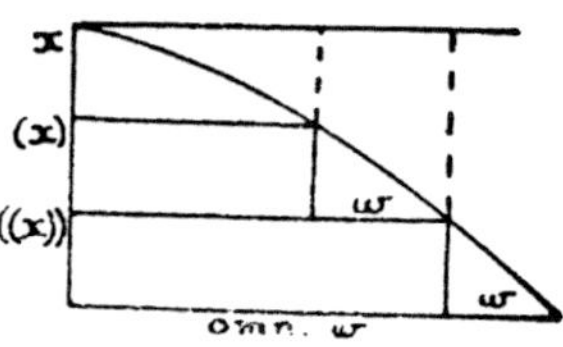

inserting this value in the preceding equation, we have

$$\text{omn.}az \sqcap \text{ult.}x^2\ \text{omn.}\frac{az}{x^2} - \text{ult.}x,\ \overline{\text{omn.omn.}\frac{az}{x^2}}\ ,$$

$$-\text{omn. ult.}x,\ \overline{\text{omn.}\frac{az}{x^2}} - \overline{\text{omn.omn.}\frac{az}{x^2}}\ ;$$

[14] I have given this equation, and those that immediately follow it, in facsimile, in order to bring out the necessity that drove Leibniz to simplify the notation.

We have here a very important bit of work. Arguing in the first instance from a single figure, Leibniz gives two general theorems in the form of moment theorems. The first is obvious on completing the rectangle in his diagram, and this is the one to which the given equation applies. In the other the whole, of which the two parts are the complements, is the moment of the completed rectangle; its equivalent is the equation

$$\text{omn.}xy = \text{ult.}x\ \text{omn.}y - \text{omn. omn.}y.$$

Now, although Leibniz does not give this equation, it is evident that he recognized the analogy between this and the one that is given; for he immediately accepts the relation as a general *analytical theorem that he can use without any reference to any figure whatever,* and proceeds to develop it further. This would therefore seem to be the point of departure that led to the Leibnizian calculus.

and this can proceed in this manner indefinitely.

$$\text{Again,} \qquad \text{omn.}\ \frac{a}{x}\ \sqcap\ x\ \text{omn.}\ \frac{a}{x^2}\ -\ \text{omn.omn.}\ \frac{a}{x^2}\ ,$$

$$\text{and} \qquad \text{omn.}a\ \sqcap\ \text{ult.}x\ \text{omn.}\ \frac{a}{x}\ -\ \text{omn. omn.}\ \frac{a}{x}\ ;$$

the last theorem expresses the sum of logarithms in terms of the known quadrature of the hyperbola.[15]

The numbers that represent the abscissae I usually call ordinals, because they express the order of the terms or ordinates. If to the square of any ordinate of a figure whose quadrature can be found, you add the square of a constant, the roots of the sum of the two squares will represent the curve of the quadratrix. Now if these roots of the sum of the two squares can also give an area that has a known quadrature, then also the curve can be rectified.[16]

[15] Having freed the matter from any reference to figures, he is able to take any value he pleases for the letters. He supposes that $z = 1$, and thus obtains the last pair of equations. He then considers x and w as the abscissa and ordinate of the rectangular hyperbola $xw = a$ (constant); hence omn.a/x or omn. w is the area under the hyperbola between two given ordinates, and therefore a logarithm; and thus omn. omn.a/x is the sum of logarithms, as he states. See Note 60, p. 122.

[16] There only seem to be two possible sources for this paragraph, (1) original work on the part of Leibniz, and (2) from Barrow. For we know that Neil's method was that of Wallis, and the method of Van Huraet used an ordinate that was proportional to the quotient of the normal by the ordinate in the original curve.

Now Barrow, in Lect. XII, § 20, has the following: "Take as you may any right-angled trapezial area (of which you have sufficient knowledge), bounded *by two parallel straight lines* AK, DL, *a straight line* AD, *and any line* KL whatever; to this let another such area be so related that when any straight line FH is drawn parallel to DL, cutting the lines AD, CE, KL in the points F, G, H, and *some determinate line* Z *is taken, the square on* FH *is equal to the squares* on FG and Z. Moreover, let the curve AIB be such that,

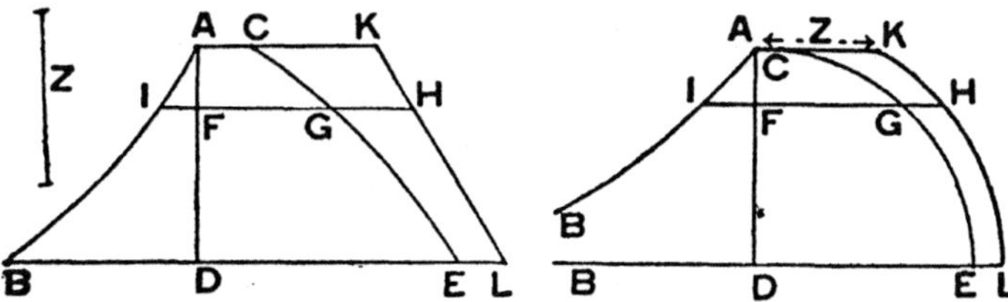

if the straight line GFI is produced to meet it, *the rectangle contained by* Z *and* FI *is equal to the space* AFGC; *then the rectangle contained by* Z *and the curve* AB *is equal to the space* ADLK. The method is just the same, even if the straight line AK is supposed to be infinite."

This striking resemblance, backed by the fact that there seems to be no connection between this theorem and the rest of the paper, that Leibniz gives no attempt at a proof, (indeed I very much doubt whether I could have made out his meaning from the original unless I had recognized Barrow's theorem) and that Leibniz gives 1675 as the date of his reading Barrow, almost forces one to conclude that this is a note on a theorem (together with an original

To describe a curve to represent a given progression.

From the square of a term of the progression, take away the square of a constant quantity; if the figure that is the quadratrix of the roots formed from the two squares is described, it will give the curve required; it does not follow that a rectifiable curve can be described.

The elements of the curve described can be expressed in many different ways. Different methods of expressing the elements of a curve may be compared with different methods of expressing a figure having similar parts with it, according as it is referred in different ways. Lastly, a solid having similar parts with a curve can thus far be expressed in many ways, and so also for a surface or figure having similar parts with the curve.

§ 6.

Three days later, Leibniz considers the possibility of being able to find the quadratrix in all cases, or when that is impossible, some curve which will serve for the quadratrix very approximately. He makes an examination of the difficulties that are likely to be met with and the means to overcome them, and he seems to be satisfied that the method can be made to do in all cases. But in the absence of an example of the method he proposes to adopt, he seems only to have been wasting his time. But this may be dismissed, for it is not here that the importance of this essay lies; it is altogether in what follows.

The rest of the essay is in the form of disjointed notes; it is just the kind of thing that any one would write *as notes while reading the works of others.* This is what I take it to be; and the works he is considering are those of

deduction therefrom by himself) which Leibniz has come across in a book that is lying before him, and that that book is Barrow's. Against it, we have the facts of the use of the word "quadratrix," not in the sense that Barrow uses it, namely as a special curve connected with the circle; that the quadratrix is one of the special curves that Barrow considers in the five examples he gives of the Differential Triangle method; and that another example of this method is the differentiation of a trigonometrical function which seems to be unknown to Leibniz.

Descartes, Sluse, Gregory St. Vincent, James Gregory and Barrow. Descartes he has already dismissed as impracticable in the manuscript of January, 1675; but there are indications that the former's method has still some influence. An incidental remark leads to the consideration of the *ductus* of Gregory St. Vincent; but these too are soon cast aside, truly because Leibniz does not quite grasp the exact meaning of Gregory. He then either remembers what he has seen in Barrow or refers to it again, for the next thing he gives is some work in connection with which he draws the characteristic triangle, *which is here for the first time, as far as* these manuscripts *go, the Barrow form and not the Pascal form.* He immediately obtains something important, namely,

$$\frac{\overline{\text{omn. } l}^2}{2} = \text{omn. } \overline{\text{omn. } l\frac{l}{a}}.$$

Noting that, in modern notation, l is dy, and a is dx, and also, since a is also supposed to be unity, that the final summation on the right-hand side is performed by "applying the successive values to the axis of x, while the summation denoted by omn.l is a straightforward summation, it follows that the equivalent of the result obtained by Leibniz is $\frac{1}{2}y^2 = \int y\, \frac{dy}{dx}\, dx.$

However, in attempting to put this theorem into words as a general theorem he makes an error; he quotes $\overline{\text{omn.}l}^2$ as the "sum of the squares" instead of the "square of the final y." This I think is simply a slip on the part of Leibniz, and not, as suggested by Gerhardt and Weissenborn, an indication that Leibniz confused $\overline{\text{omn.}l}^2$ with $\overline{\text{omn.}l^2}$, and considered them as equivalent. Neither of these authorities appears to have noticed the fact that when Leibniz has invented the sign $\int$ (which he immediately proceeds to do) he carefully makes the distinction between the equivalents to the square of a sum and the sum of the squares. Thus we find that his equation is written as

$$\int \overline{\frac{l^2}{2}} = \int \int \overline{l \frac{l}{a}},$$ (note the vinculum)

while later in the essay we have $\int l^3$ to stand for the sum of the cubes. Further, apart from this. I do not think that any one can impute such confusion of ideas to Leibniz, if it is noted that so far this is not the differential calculus, but the calculus of differences, i. e., l is still a very small but finite line and not an infinitesimal; for in § 4 Leibniz had squared a trinomial successfully, and must have known that the sum of the squares could not be equal to the square of the sum. Both these above-named authorities seem to find some difficulty over the introduction of the letter a, apparently haphazard. This difficulty becomes non-existent, if it is remembered that a is taken to be unity, and the remarks made about dimensions by Leibniz are carefully considered; it will then be found that the a is introduced to keep the equations homogeneous! Weissenborn also remarks that Leibniz jots down the integral of x^2 without giving a proof, and appears to be in doubt how he reached it. If this is so, it confirms the opinion that I have already formed, namely, that neither Gerhardt nor Weissenborn tried to get to the bottom of these manuscripts, being content with simply "skimming the cream."

I suggest that Barrow, Gregory St. Vincent, and even Sluse, now join Descartes on the shelf or the floor, and that the rest of the essay is all Leibniz. He writes the two equations he has found, the equivalents to two theorems obtained geometrically, notes the fact that these are true for infinitely small differences (without, however, mentioning that they are *only* true in such a case), discards diagrams, and proceeds analytically; that is, the y's are successive values of some function of x, where the values of x are in arithmetical progression; hence, substituting x for l in the equation

$$\text{omn.}xl = \text{omn.}l - \text{omn.omn.}l,$$

and remembering that omn.$x = x^2/2$, as he has proved, we have

$$\text{omn.}\,x^2 = x\,\frac{x^2}{2} - \text{omn.}\frac{x^2}{2}, \text{ or omn.}\,x^2 = \frac{x^3}{3}.$$

Again, below he gives $\int x^3 = \dfrac{x^4}{4}$ correctly (although there is an obvious slip or, as I think, a misprint of l for x): this could have been obtained in the same way.

$$\text{omn.}\,x^3 = x\frac{x^3}{3} - \text{omn.}\frac{x^3}{3}, \text{ or omn.}\,x^3 = \frac{x^4}{4}.$$

Similarly, Leibniz could have gone on indefinitely, and thus obtained the integrals of all the powers of x. But his brain is too active; as Weissenborn says, his soul is in the throes of creation. He merely alludes in passing to the inverse operation to $\int$ as being represented by d, which he for some reason writes in the denominator (probably erroneously because he has noted that $\int$ increases the dimensions); and then he harks back to the opening idea of the essay, the obtaining of the quadratrix by means of transformation of equations, an idea truly as hopeless as the method of Descartes which he has discarded. Nevertheless, even then he obtains something remarkable, nothing more or less than the inverse of the differentiation of a product. This fundamental theorem is obtained geometrically; the proof of the little theorem on which the final result is founded is not given, neither is there a diagram. It cannot therefore be supposed but that Leibniz is working from a diagram already drawn, and I suggest he was referring to one of those theorems, with which he had filled "hundreds of pages" between 1673 and 1675. The proof follows quite easily by the use of the characteristic triangle, and is given in a footnote. This theorem is not in Barrow, nor can I remember seeing it in Cavalieri; I have not yet been able to procure a Gregory St. Vincent; it may be in James Gregory.

The benefits of this discovery are lost as before, for Leibniz once more alludes to the transformation of equations for the purpose of obtaining the quadratrix.

Summing the whole essay, we can say that in it is the beginning of the Leibnizian *analytical calculus.*

29 October, 1675.

Analyseos Tetragonisticae pars secunda.

[Second part of analytical quadrature.]

I think that now at last we can give a method, by which the analytical quadratrix may be found for any analytical figure, whenever that is possible; and, when it can not be done, it will yet always be possible that an analytical figure may be described, which will act as the quadratrix as nearly as is required. This is how I look at it:

Suppose the equation of the curve, of which the quadratrix is required, is given, and that the unknowns in it are x and v. Let the equation to the curve required be[17]

$$v = b + cx + dy + ex^2 + fy^2 + gyx + hy^3 + lx^3 + mxyy + yxx + \text{etc.} ; \quad (i)$$

let it be set in order for tangents, as follows:

$$- dy - 2fy^2 - gyx - 3hy^3 - 2mxy^2 - mx^2y - \text{etc.}$$
$$= ct + 2ext + gyt + 3lx^2t + my^2t + 2yxt + \text{etc.} \ldots\ldots\ldots (ii)$$

[17] This is either a misprint, v instead of O, or else Leibniz is in error. For Slusius's method there must be only two variables in the equation. In the *Phil. Trans.* for 1672 (No. 90), Sluse gives his method thus:

If $y^5 + by^4 = 2qqv^3 - yyv^3$, then the equation must be written $y^5 + by^4 + yyv^3 = 2qqv^3 - yyv^3$; then multiply each term on the left-hand side by the number of y's in the term, and substitute t in place of one y in each; similarly multiply each term on the left-hand side by the exponent of v; the equation obtained will give the value of t.

The use of the letters v and y is to be noted in connection with Leibniz's use of the same letters; it does not seem at all necessary that Leibniz should have seen Newton's work, with this ready to the former's hand, as a member of the Royal Society. I suggest that Sluse obtained his rule by the use of a and e, as given in Barrow. Can Barrow's words *usitatum a nobis* (in the midst of a passage written in the first person singular) have meant that the method was common property to himself *and several other mathematicians that were contemporary with him?* This would explain a great deal.

Now, $t/y = a/v$; hence, if from the equation $t/y = a/v$, we eliminate t and y by the help of equations (i) and (ii), that equation should be produced which represents the figure that has to be quadratured; and by comparing the terms of the equation thus obtained with the given equation, unless indeed there is no possibility of comparing them, we shall obtain the quadrature.

But if an impossibility arises, it is then known that the given analytical figure has no analytical quadratrix. But it is quite clear that if we add to it such as will change it almost imperceptibly, then a quadrable figure may be obtained, since this plainly produces another equation. However, as an impossible case may arise, we must consider the difficulties.

Say that the equation that is obtained is of infinite prolixity, while the given one is finite. My answer is, that in comparing the one with the other it will be seen how far at most the powers of the unknowns in the indefinite equation can go. The retort may be made, that it may happen that the indefinite equation obtained may have more terms than the finite equation that is given and yet may be reduced to it, for it may be divided by something else that is either finite or indefinite. This difficulty hindered me for a long time a year ago, but now I see that we should not be stopped by it. For it may happen that from a certain determinate figure (whose equation is not divisible by a rational) by the method of tangents there may arise an ambiguous figure; for it is impossible to say that, for *any* figure, there shall be only one tangent at any one point. Hence the produced equation can neither be divided by a finite nor by an indefinite quantity; for in truth indefinite figures, or those whose ordinates are represented by an infinite equation, have sometimes these very ordinates finite, and these ought to satisfy the equation. Notwithstanding that, I foresee another difficulty; for indeed it seems that sometimes it may happen that all the roots of the equation will not serve for the solution of the problem. Yet, to tell the truth, I believe they will do so.

Now here is a difficulty that really is great. It may happen that a finite equation may also be expressed as an indefinite one, so that the equation obtained may really be the same as the given equation although it does not appear to be. For example,

$$y^2 = x/(1+x) = x - x^2 + x^3 - x^4 + x^5 - x^6 + \text{etc.} ;$$

and in the same way others can be formed by various compositions and divisions. This I confess is truly a difficult point, but it can be answered thus: If a figure has an analytical quadratrix of any sort, in all cases it may be assumed to be an indefinite one; and then it will not in all cases give an indefinite, but sometimes a finite, equation that is equivalent to the given equation. In the same way, it is certain that the quadratrix of a given curve as it is usually investigated, whenever there is one, will also be determined; and that too given uniquely and not ambiguously, so that any that differs from it, differs only in name. There is still one difficulty left; it seems impossible to determine which is the end or first term of the indefinite equation that is obtained; for it may happen that the terms of lower degree are cut out, and then it is divisible by y or x or yx or powers of these; nor do I see that there is anything to prevent this. There is the same difficulty whether you start from the lowest or the highest degree in the equation assumed to begin with as indefinite. Suppose then that in the equation obtained this

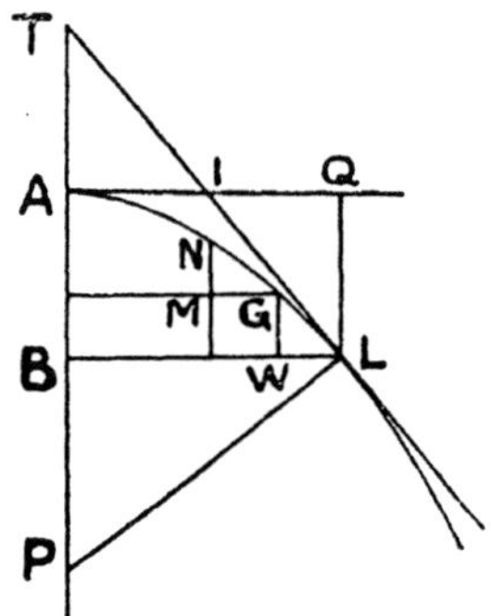

division is possible, then it is necessary that the constant term should be absent, and also all those terms in which x alone or, if you like, all the terms in which y alone is absent; and if we examine this continuously we may light upon an impossibility.

In this general calculus then, we may take it as certain that this difficulty is solved, and that such a division after the calculation can never happen; or if it is possible for it to happen, then the terms will go out, one after the other, so that the equation can be depressed and the comparison be made; and then it is to be seen whether this difficulty cannot be overcome in general, and the comparison proceed as we proceed with the elimination. Perhaps if the figure to be quadratured is reduced beforehand to its simplest equation possible, impossibilities will the more readily be detected. For then presumably the quadratrix must become more simplified. In addition we have another source of assistance; for various calculations leading to the same thing, though obviously differing from one another, can be contrived, from which equations are comparable.

Let BL$=y$, WL$=l$, BP$=p$, TB$=t$, GW$=a$, then $y=$omn.l.

Incidentally I may remark that there are composite numbers that cannot be added or subtracted from one another by parts, namely those denominated by powers, or by sub-powers or surds. There are also other denominate numbers which cannot be multiplied together by parts, such as numbers representing sums; for instance, omn.l cannot be multiplied by omn.p, nor can we have $y^3=2$omn. omn.pl. However, as such a multiplication may be imagined to occur under certain conditions, we must consider it as follows:

We require the space that represents the product of all the p's into all the l's; we cannot make use of the ductions of Gregory St. Vincent, where figures are multiplied by figures, for by this method one ordinate is not multiplied by all the others, but one into one. You may say that if one ordinate is multiplied by all the rest it will produce a sursolid space, namely, the sum of an infinite number of solids. For this difficulty I have found a remedy that is really admirable. Let every l be represented by an infinitely short straight line WL, that is, we want the quadratrix line representing omn. l; well, the line BL$=$omn. l; and if this is multiplied by every p, each represented by a plane figure, then a solid is produced. If all the l's are straight lines and all the p's are curves, a curved surface is produced by a duction of the same sort. But these things are all old; now, here is something new.

If upon WL, MG, or every single l, is superimposed the same curve representing all the p's, where the curve p is originally all in the same plane and is carried along the curve AGL while its plane always moves parallel to itself, then what we require will be obtained. In place of a curve a plane may be carried along the curve in the same manner, and a solid will be obtained, whereas by the former method it was a curvilinear surface; and both for the surface and for the solid the section always remains the same. It remains to be seen whether a number of analytical surfaces cannot be ascertained, as in the case of analytical lines; but this is mentioned only incidentally.

N. B. The curvilinear surface formed by the motion of a curve parallel to itself along the curve will be equal to the cylinder

of the curve under BL, the sum of all the l's but this is also mentioned incidentally.

To resume, $\dfrac{l}{a} = \dfrac{p}{\text{omn.}\, l} = y$, therefore $p = \dfrac{\overline{\text{omn.}\, l}}{a}\, l$.   Hence,

omn. $y\, \dfrac{l}{a}$ does not mean the same thing as omn.y into omn.l, nor yet

y into omn.l; for, since $p = \dfrac{y}{a}\, l$ or $\dfrac{\overline{\text{omn.}\, l}}{a}\, l$, it means the same thing

as omn.l multiplied by that one l that corresponds with a certain

p; hence, omn.$p = $ omn.$\dfrac{\overline{\text{omn.}\, l}}{a}\, l$.   Now I have otherwise proved

omn.$p = \dfrac{y^2}{2}$, i. e., $= \dfrac{\overline{\text{omn.}\, l}^{2}}{2}$; therefore we have a theorem that to me

seems admirable, and one that will be of great service to this new
calculus, namely,

$$\frac{\overline{\text{omn.}\, l}^{2}}{2} = \text{omn.}\ \overline{\text{omn.}\, l}\,\frac{l}{a}\,, \quad \text{whatever } l \text{ may be;}$$

that is, if all the l's are multiplied by their last, and so on as often
as it can be done, the sum of all these products will be equal to half
the sum of the squares, of which the sides are the sum of the l's
or all the l's.   This is a very fine theorem, and one that is not at all
obvious.

Another theorem of the same kind is:

$$\text{omn.}\,xl = x\,\text{omn.}\,l - \text{omn.omn.}\,l\,,$$

where l is taken to be a term of a progression, and x is the number
which expresses the position or order of the l corresponding to it;
or x is the ordinal number and l is the ordered thing.

N. B.   In these calculations a law governing things of the same
kind can be noted; for, if omn. is prefixed to a number or ratio, or
to something indefinitely small, then a line is produced, also if to
a line, then a surface, or if to a surface, then a solid; and so on
to infinity for higher dimensions.

It will be useful to write $\int$ for omn., so that

$$\int l = \text{omn.}\,l,\ \text{or the sum of the } l\text{'s.}$$

Thus,
$$\int \frac{l^2}{2} = \int \overline{\int l\, \frac{l}{a}}\,, \quad \text{and}\ \int \overline{xl} = x \int \overline{l} - \int \int l.$$

From this it will appear that a law of things of the same kind should always be noted, as it is useful in obviating errors of calculation.

N. B. If $\int l$ is given analytically, then l is also given; therefore if $\int \int l$ is given, so also is l; but if l is given, $\int l$ is not given as well. In all cases $\int x = x^2/2$.

N. B. All these theorems are true for series in which the differences of the terms bear to the terms themselves a ratio that is less than any assignable quantity.

$$\int x^2 = \frac{x^3}{3}$$

Now note that if the terms are affected, the sum is also affected in the same way, such being a general rule; for example,

$$\int \overline{\frac{a}{b} l} = \frac{a}{b} \times \int l \,,$$ that is to say, if $\frac{a}{b}$ is a constant term, it is to be multiplied by the maximum ordinal; but if it is not a constant term, then it is impossible to deal with it, unless it can be reduced to terms in l, or whenever it can be reduced to a common quantity, such as an ordinal.

N. B. As often as in the tetragonistic equation, only one letter, say l, varies, it can be considered to be a constant term, and $\int l$ will equal x. Also on this fundamental there depends the theorem:

$$\int \frac{l^2}{2} = \int \int \overline{l\,l} \,, \text{ that is, } \frac{x^2}{2} = \int x.$$

Hence, in the same way we can immediately solve innumerable things like this; thus, we require to know what e is, where

$$\int \frac{c}{a} \overline{\int l} + ba^2 + \int l^3 \overset{(18)}{+} \int l^3 = ea^3;$$

we have

$$a^3 e = \frac{cx^3}{3} + ba^2 x + \frac{x^4}{4} + xa^3.$$

For indeed $\int l^3 = x$, because l is supposed to be equal[19] to a for the purpose of the calculation; $\int \frac{l}{a} = x.$

[18] There is evidently a slip here; l should be x.

[19] This is an instance of the care which Leibniz takes; in the work above l has been the difference for y, and a the difference for x; he is now integrating an algebraical expression, and not considering a figure at all; hence $l = a$, and a is equal to unity, and therefore $\int l^3 = l^3 x = a^3 x = x!$ Thus what is generally considered to be a muddle turns out to be quite correct. The muddle is not with Leibniz, it is with the transcriber. It is certain that these manuscripts want careful republishing from the originals; won't some millionaire pay to have them reproduced photographically in an *edition de luxe*?

Also $\qquad \int c \int \overline{l^2} = \dfrac{cx^3}{3}$, that is $= \dfrac{c \int \overline{l^3}}{3a^3}$, $\qquad \int ba^2 = \int l\, ba.$

Also it is understood that a is unity. These are sufficiently new and notable, since they will lead to a new calculus.

I propose to return to former considerations.

Given l, and its relation to x, to find $\int l$.*

This is to be obtained from the contrary calculus, that is to say, suppose that $\int l = ya$. Let $l = ya/d$; then just as $\int$ will increase, so d will diminish the dimensions. But $\int$ means a sum, and d a difference. From the given y, we can always find y/d or l, that is, the difference of the y's. Hence one equation may be transformed into

the other; just as from the equation $\int c \int \overline{l^2} = \dfrac{c \int \overline{l^3}}{3a^3}$, we can ob-

tain the equation $c \int \overline{l^2} = \dfrac{c \int \overline{l^3}}{3a^3 d}$.

N. B. $\displaystyle\int \frac{x^3}{b} + \int \frac{x^2 a}{e} = \int \frac{x^3 + x^2 a}{b \qquad e}$. And similarly,

$$\frac{x^3}{db} + \frac{x^2 a}{de} = \frac{\dfrac{x^3}{b} + \dfrac{x^2 a}{e}}{d}.$$

But to return to what has been done above. We can investigate $\int l$ in two ways; one, by summing y and seeking $ya/d = l$, the other, by summing $z^2/2a = y$, or by summing $\sqrt{2ay} = z$, and then $z^2/t = p = l = ya/d$. Hence, if in an indefinite equation, we eliminate y by substituting in its place $z^2/2a$, and investigate the t of this new equation which is indefinite like the first, and then by the help of the value $z^2/t = l$, and after that by the help of the new value of t, eliminate z from the indefinite equation containing z and t, there will remain out of the (three) letters x, z, t, l, the letter l alone; and again we ought to get an equation which should be the same not only as the given one, but also the same as the one that was obtained a little while ago. Hence, since we have two indefinite equations, containing not only the principle quantities, but also arbitrary ones, yet not altogether unlike the former; and these ought to be identical; it will appear to show whether certain terms cannot be eliminated, whether it is not possible that a comparison should be made, and other things of the sort; and, what

---

* This is, as I am going to show later, on p. 180, palpably a mere analytical translation of Barrow's geometry.

is really the most important thing, which terms are really the greatest and the least, or the number of terms of the equation.

Moreover, since in the similar triangles TBL, GWL, LBP, no mention has yet been made of the abscissa x or of the fixed point A, let us then suppose that through the fixed point A there is drawn an unlimited straight line AIQ, parallel to LB, meeting the tangent LT in I; and let AQ=BL; bisect AI in N; then I say that the sum of every QN will always be equal to the triangle ABL, as can easily be shown by what I have said in another place.[20]

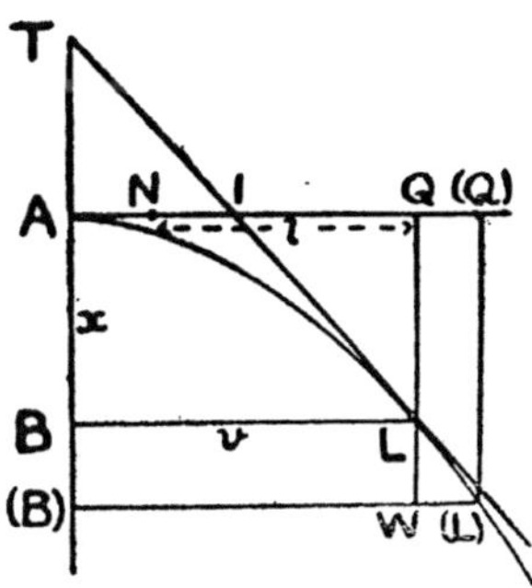

These considerations give once more a fresh fundamental theorem for the calculus. For $xv/2=y$, where we suppose that BL$=v$ and QN$=l$, and $y=\int l$;

but
$$\frac{AI}{v} = \frac{t-x}{t}, \quad \text{therefore} \quad AI = \frac{t-x}{t}\,v,$$

and
$$QI = v - AI = v\frac{tt}{t}v + \frac{xv}{2t}, \quad \text{i. e. } QI = \frac{xv}{t},$$

$$\therefore QN = QI + \frac{AI}{2} = \frac{xv}{t} + \frac{v}{2} - \frac{xv}{2t} = \frac{xv + tv}{2t} = l. \tag{21}$$

Now, by the help of the equation $(xv+tv)/2t=l$, and of the former equation $y=xv/2$, and taking once more the first indefinite or general equation as a third, and eliminating first of all y, then t by means of the value found for the ratio of t to x from the indefinite equation containing x and v, and lastly v by the help of the equation $(xv+tv)/2t=l$, in which the principal quantities x and l alone remain, as before; and this again should be identical with the given equation.

Thus we have found three equations obtained in different ways, which should all be identical with one another and with the given equation; and these three are not only identical but should also consist of the same letters and signs; and whether this is possible, will immediately appear on being worked out analytically.

[20] Since the triangles QLI, WL(L) are similar, QI.B(B)$=$QL.Q(Q), hence omn.QI (applied to AB)$=$omn. QL (to AQ)$=$figure AQLA, hence omn.(QI$+$QA)$=$rect. ABLQ$=2\triangle$ABL.

[21] Since l is the difference for y, therefore $2l$ is the difference for xv; this is shown to be $(xv+tv)/t$ or $x(v/t)+v$; and this is the equivalent to (since $v/t=dv/dx=dv$)
$$d(xv) = xdv + v = xdv + vdx.$$

§ 7.

The next manuscript is a further continuation of the preceding, written two days later. In this Leibniz returns to the idea that he has found so prolific, namely, the moments of a figure. It is to be observed that he speaks of the method of breaking up an area into segments as something that he has already worked out; this will be remarked upon in a note on a later manuscript, where it will help to clear up a small difficulty. The accuracy of the rather involved algebraical work is also a point to be noticed.

1 November, 1675.

Analyseos Tetragonisticae pars tertia.

[Third part of analytical quadrature.]

It was some time ago that I observed that, being given the moment of a curve ABC, or of a curvilinear figure DABCE, about two straight lines parallel to one another, such as GF, LH (or MN, PQ), then the area of the figure could be obtained; because the two moments differed from one another by the cylinder of the figure, where the altitude was the distance between the parallels.

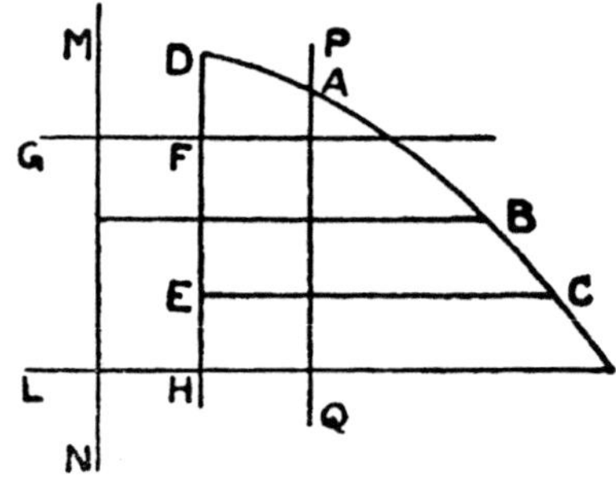

Now, this is true of every progression, whether of numbers or of lines; that is, even if we do not use curvilinear figures but ordinated polygons; in other words, where the differences between the terms are not infinitely small. Suppose we have any such ordinated quantity z and let the ordinal number be x, then

$$b \overline{\mathrm{omn}.z} \sqcap \pm \mathrm{omn}.\overline{zx} \mp \mathrm{omn}.\overline{zx+b}$$

and this is evident by the calculus alone.

By this rule, can be found the sums of terms of an arithmetical progression refolded reciprocally;[22] and this multiplication takes

---

[22] The meaning of this is probably a series such as that considered by Wallis. If a, $a + d$, $a + 2d$, etc. is the arithmetical progression, and l, $l — d$, $l — 2d$, etc. is the series reversed, then the series refolded reciprocally is al, $(a + d)(l — d)$, $(a + 2d)(l — 2d)$, etc. It may however mean the sum of the squares of the arithmetical progression. But the point is not very important.

place when it is required to find the moment of the ordinates about a straight line perpendicular to the axis. But if the moment about any other straight line is required, there is the following general rule:

From the center of gravity of each of the quantities of which the moment is required, a perpendicular is drawn to the axis of libration; then the sum of the rectangles contained by the distances or perpendiculars and the quantities will be equal to the moment about the given straight line.

Hence, if the given straight line is the axis of equilibrium, it immediately follows that the moment of the figure about the axis is equal to the sum of the half-squares. Also when it is parallel to that, it will differ from the foregoing by a known quantity.

Now, let us take another straight line: for the circle for instance, let ABCD be a quadrant, vertex A, and center D; let another straight line be given, that is to say, let the prependicular DF be given and

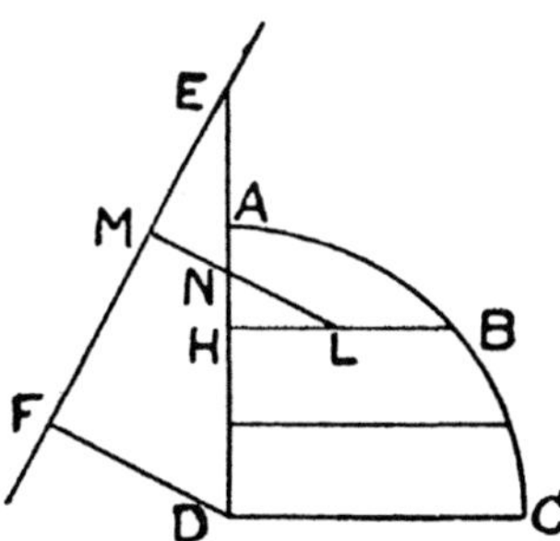

also EF where it meets the diameter, and thus also DE; let HB be the general ordinate to the circle, and L its middle point; let LM be drawn perpendicular to EF.

Then it is clear that the triangles EFD, EMN (where N is the intersection of ML and AD), and LHN are similar.

Let $AD = x$, then $HL = \dfrac{y}{2} = \dfrac{\sqrt{a^2 - x^2}}{2}$. But, on account of

the similar triangles, $\dfrac{NH}{HL} = \dfrac{DF\,(=d)}{FE\,(=f)}$;

therefore

$$NH = \frac{d}{2f}\sqrt{a^2 - x^2} = \frac{yd}{2f}.$$

Hence, $\text{EN} = \text{DE}(=e) - \text{HD}(=x) - \text{NH}\left(=\dfrac{yd}{2f}\right) = e - x - \dfrac{yd}{2f}$.

Now $\text{NL} = \sqrt{\text{NH}^2 + \text{HL}^2} = \sqrt{\dfrac{d^2}{4f^2}y^2 + \dfrac{y^2}{4}} = \dfrac{y}{2}\sqrt{\dfrac{d^2}{f^2}+1}$;

and $\dfrac{\text{MN}}{\text{EN}} = \dfrac{\text{NH}}{\text{HL}}$, or $\text{MN} = \dfrac{\text{NH.EN}}{\text{HL}}$; thus we have

$$\text{MN} = \dfrac{dy}{2f}\cdot\dfrac{e-x-\dfrac{dy}{2f}}{\dfrac{y}{2}\sqrt{\dfrac{d^2}{f^2}+1}} = \dfrac{d}{f\sqrt{\dfrac{d^2}{f^2}+1}}\;\overline{e-x-\dfrac{dy}{2f}}\,;$$

and $\text{ML} = \text{MN} + \text{NL} = \dfrac{d}{f\sqrt{\dfrac{d^2}{f^2}+1}}\;\overline{e-x-\dfrac{dy}{2f}}+\dfrac{y}{2}\sqrt{\dfrac{d^2}{f^2}+1}$;

hence, since $e = \sqrt{f^2-d^2}$, we have [23]

$$\text{ML} = \dfrac{d\sqrt{f^2-d^2}-x-\dfrac{d}{2f}y+\dfrac{d^2+f^2}{2f}y}{\sqrt{d^2+f^2}} = \dfrac{d\sqrt{f^2-d^2}-x+\dfrac{fy}{2}}{\sqrt{d^2+f^2}}$$

and this calculation is general for any curve, so long as x is always taken as the abscissa and y as the ordinate.

Therefore the rectangle contained by ML and HB $(=y)$, or the moment of each ordinate taken with regard to the straight line EF, or wa, will be equal to

$$\dfrac{d\sqrt{f^2-d^2}\,y - xy + \dfrac{f}{2}y^2}{\sqrt{f^2+d^2}}$$

Hence, omn.w will be obtained from the known values of omn.x, omn.xy, and omn.y^2; also, if any three of these four are given, the fourth is also known.

Now, omn.xy will be equal to the moment of the figure about the vertex, omn. y^2 will be equal to the moment of the figure about the axis; hence, given three moments of the figure, that is to say, the moments about two straight lines at right angles and any third, the area is given.

This theorem, however, is less general than the one that was given before, in the first part of this essay, where it does not matter

<hr>

[23] The accuracy of the algebra is noteworthy in comparison with the inaccuracies that occur later. There is however a slip: $e^2 = f^2 + d^2$ and not $f^2 - d^2$; this must be a slip and not a misprint, because it persists throughout. It should be noted that the figure given by Gerhardt is careless in that LM is made to pass through A.

what the angle between the straight lines may be, if only we are given three moments; but it is always understood that they are in the same plane. (Meanwhile, however, this theorem will suffice for the curve of the primary hyperbola; for, if f is infinite, or if FE and ED are parallel, $dy + y^2/2 = wa$, as has already been proved.)

It is to be observed that by other calculation the area of a quantity, whose center of gravity lies in a given plane (even though the whole quantity does not), can be found from three given moments about three straight lines in that plane. From this it is to be seen whether the results obtained, when compared with one another, will not produce something new.

If instead of the moment of a figure we require the moment of all the arcs BP, PC, etc., the perpendiculars are to be drawn from the points B, P, C, etc. only, to the straight line; for it will make no difference whether they are drawn from the end or from the middle of BP, for instance, for the difference between two such perpendiculars is infinitely small. Hence, calling the element of the curve z, the moment of the curve about the straight line EF is

$$\frac{d\sqrt{f^2 - d^2}\,z - dxz + fyz}{\sqrt{d^2 + f^2}}$$

Most of the theorems of the geometry of indivisibles which are to be found in the works of Cavalieri, Vincent, Wallis, Gregory and Barrow, are immediately evident from the calculus; as, for instance, that the perpendiculars to the axis are equal to the surface or moment of the curve about the axis, for you find that a perpendicular is equal to the rectangle contained by an element of the curve and the ordinate. Therefore I do not set any value on such theorems, or on those about applications of intercepts on the axis (intercepted between the tangents and the ordinates) to the base. Such theorems bring forth nothing new, except maybe they afford formulas for the calculus.

But my theorem about the dimensions of the segments does bring out a new thing, because the space whose dimension is sought is broken up in a different way, that is to say, not only into ordinates but into triangles. Also perhaps the Centrobaric method yields something new. Maybe an easy method can be obtained by which, without diagrams, those things which depend on a figure can be derived by calculus. Gregory's theorem, on ductions of two parabolas,[24] one under the other, equal to a cylinder, is immediately evident by calculus; for the ordinate of a circle $y = \sqrt{a^2 - x^2}$, that is, the product of $\sqrt{a+x}$ and $\sqrt{a-x}$; and in the same way, $\sqrt{2av - v^2} = y$, which gives $y = \sqrt{v}$ into $\sqrt{2a - v}$; and these come to the same thing.

If the same ordinate y is multiplied by some quantity z, and afterward by the same $z \pm$ some known or constant number b, the difference between the sums produced will be equal to the cylinder of the figure; so that

$$zy,, - zy + by \sqcap by.$$

Although this is evident in general by itself, yet applications of it are not always evident. For instance, let

$$y = \frac{x^2}{ax - b^2} = \frac{x^2}{\sqrt{ax + b}, \ \sqrt{ax - b}} ;$$

then, multiplying by $\sqrt{ax + b}$, we have $\dfrac{x^2}{\sqrt{ax - b}}$; $\ldots\ldots\ldots\ldots$ (A)

and, multiplying by $\sqrt{ax - b}$, we have $\dfrac{x^2}{\sqrt{ax + b}}$; $\ldots\ldots\ldots$ (B)

but, since instead of $\dfrac{ax^2}{ax - b^2}$, we can have $x + \dfrac{b^2 x}{ax - b^2}$,

which depends on the quadrature of the hyperbola; and thus if one of the two things, (A) or (B), is given, then the other is also known, supposing that the quadrature of the hyperbola is known.

Suppose that at the points C, D, E of a curve situated in any plane there are imposed, perpendicular to the plane, the ordinates of another curve FGH (not necessarily of the same constitution), in such a manner that the middle point of each of these ordinates lies in the plane; then it is evident that LC, MD, NE, multiplied by FL, GM, HN, (that is, the lines imposed at C, D, E of the curve BCDE) or the rectangles FLG, GMD, HNE, or the duction of these two planes into one another, will be equal to the moment of every LC, MD, NE, etc. Hence, if PR is another axis, and the interval between it and QL is the straight line PQ, the moment

<hr>

[24] Such theorems are also considered in Wallis, where it is shown that the products for two equal parabolas are the squares on the ordinates of a semicircle; the axes of the parabolas being coincident, but set in opposite sense.

about PR differs from that about QL by the cylinder whose base is LC, MD, etc., and whose height is PQ.[25]

But, if we know the moment about the straight line PQ, and also that about some other straight line in another position, as TS, of all

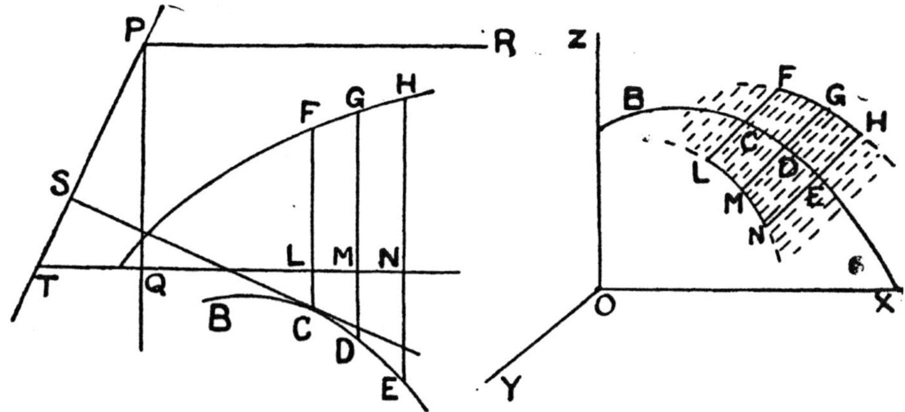

the ordinates LF of the same figure, imposed at the points C, then we shall have the cylinder corresponding to all the LF's, as I will now prove.

If we call QL, x, and CL, y, then TC $= \dfrac{f}{a}x + \dfrac{g}{a}y + h$; and this multiplied by z, where FL or MG$=z$, will give

$$\frac{f}{a}xz + \frac{g}{a}yz + hz.$$

Now xz is given, being the supposed moment about PQ, which is the same whether the z's are placed where they were in the lines LF, MG, etc., or at the points C, D, E. Also yz is given, either as the rectangle FLC or as the duction, by hypothesis. Hence, if in

[25] This is obviously wrong; the base of the cylinder is the area made up of FL, GM, HN, etc. The whole of this last passage proved to be difficult to make out; Leibniz has not completed his figure, by showing the surface formed by placing the ordinates FL, GM, HN with their middle points at C, D, E, and the ordinates themselves perpendicular to the plane of the curve BCDE, which figure I have added on the right-hand side of Leibniz's figure. Even when this is given, there is another difficulty added because as given by Gerhardt, CS is the tangent at D instead of the proper line, namely, the perpendicular from C to TS; in addition through a misprint, this line is afterward referred to as TC. Lastly, "the rectangle FLG" is a misprint for FLC, which with Leibniz stands for FL.LC; this notation for a rectangle is, as far as I can remember, used by Wallis and Cavalieri.

When all these errors are revised, what at first sight seemed to be rather a muddle turns out to be an exceedingly neat idea in connection with the moments of a figure, and their use to find an area, although mostly impracticable; it is evidently taken from Pascal (cf. onglets).

Note. The values f, g, a, h, are the lengths of TQ, QP, PT, and the perpendicular from Q on PT.

addition there is given one moment of the ordinates imposed upon the curve at the points C, D, E, and this is taken to be equal to $\frac{f}{a} xz + \frac{g}{a} yz + hz$, then we have hz or the cylinder required.

Hence, the curve BCDE is to be chosen such that the ordinates of the given curve can be multiplied by different ordinates of the former, drawn either to the axis QL or to the axis TS, with some advantage of simplicity; and the curves that are suitable for this are those that have several suitable axes, such as the circular or primary hyperbola, which has a pair of asymptotes, or an axis and a conjugate axis.

§ 8.

Much comment has been made on the fact that the date of the next manuscript was originally "11 November 1675"; that the 5 had been altered to a 3, the ink being of a darker shade; and that it is almost certain that this alteration in date was made for some ulterior motive by Leibniz himself. Hence, if he was capable of falsifying a date in one particular case, then he is not to be trusted in others,...., and so on. Instead of trying to explain away this alteration, let us try to find an explanation as to the reason of its having been made by Leibniz; I offer the following as at least feasible.

The essay starts with the words, *"Jam superiore anno mihi proposueram questionem,...."* I suppose that by this Leibniz intended: "A year or two ago, I set myself the question,....." This conforms with what follows; the theorem that he sets down is one such as those that were suggested to him by Huygens, and further theorems that came to him as deductions during his first intercourse with Huygens. Years later, I therefore suggest, Leibniz refers to this manuscript, reads his own Latin, *superiore anno,* as "in the above year," gets no further, recognizes the theorem by its figure as one of the Huygens-time batch, and says to himself "1675? No, that's wrong, should be 1673," and proceeds to alter it to what he remembers was the date for the first consideration of the theorem.

N. B. Gerhardt himself has remarked on the darker tint of the ink used in the alteration; hence my argument, made at a later date.

The date 1675 is incontestable; for this composition is quite glaringly a development of the work that has been so efficiently started in that of November 1, 1675. Progress is still delayed by the idea that has obsessed Leibniz up till now, that of the transformation of equations, so as to be able to eliminate more unknowns than the original number of his equations warrant. He sets himself the problem: "To determine the curve in which the distance between the vertex and the foot of the normal is reciprocally proportional to the ordinate," i. e., the solution of the equation $x + y\, dy/dx = a^2/y$, in modern notation. This is a very unlucky choice for him: for I have it on the authority of Prof. A. R. Forsyth that this is incapable of solution in ordinary functions or even by a series in which the law of the series is easily and simply expressible—at least he confesses that he is unable to obtain such a solution, which I take it comes to the same thing.

Leibniz professes to have found the solution and gives $(y^2 + x^2)(a^2 - yx) = 2y^2 \log y$; and unfortunately this false success but enhances the value in his eyes of the method mentioned above. But from the equation given as the solution we may draw an incontestable conclusion; for in a previous problem Leibniz verifies his solution by the method of tangents, i. e., by differentiation, although the method does not as yet convey that idea to him; but he does not verify the solution in this case, *because he is unable at this date to differentiate the product* $y^2 \log y$.

The introduction of dx instead of x/d marks a further advance, more important perhaps than the use of $\int y\, dy$;

for he still writes $\int x$, considering dx to be constant and equal to unity. He is beginning to grasp the infinitesimal nature of his calculus, and that infinitesimals are not to be neglected because of their intrinsic smallness, but because of their smallness *with respect to other quantities* which come into the same equations and are finite; but he is far from being certain about it as yet, as is evidenced by the discussion as to whether $d(v/\psi) = dv/d\psi$ or not. However, the whole manuscript marks a distinct advance on anything that has gone before. From now on he probably discards geometry, and refers to Descartes, Gregory and Barrow only for examples to show how much superior is his method to theirs. I put his final reading of Barrow down to the interval between the date of this manuscript, 11 November, 1675, and November, 1676; it is at this time that he inserts his sign of integration in the margins of the theorems. The next person that examines the originals of these manuscripts (I am convinced that this is very necessary), should carefully see whether the **ink used** for the note "*novi dudum*" (which I have mentioned) is the same as that used for the sign of integration; also the other books that were used by Leibniz in his self-education should be searchingly scrutinized for clues.

The last remark I have to make is one of astonishment at the errors in the algebraical work which brings this essay to a close, and to a less degree throughout the essay; for we have seen the accuracy to which Leibniz has attained in a previous manuscript; of course, a great deal of erroneous work can be explained by supposing none too careful transcription; but a re-examination of the whole of the Leibnizian remains should include a careful scrutiny on the point as to whether some of the extracts given by Gerhardt are not the work of pupils of Leibniz, whose writing would naturally be somewhat similar. Perhaps too some of those early geometrical theorems might be unearthed; and this would well reward the most painstaking search. Nobody can assert that anything like an adequate tale of the progress of the Leibnizian genius has so far been told.

11 November, 1673.[26]

Methodi tangentium inversae exempla.

[Examples of the inverse method of tangents.]

A year or two ago I asked myself the question, what can be considered one of the most difficult things in the whole of geometry, or, in other words, what was there for which the ordinary methods had contributed nothing profitable. To-day I found the answer to it, and I now give the analysis of it.

Find the curve C(C), *in which* BP, *the interval between the ordinate* BC *and* PC *the normal to the curve, taken along the axis* AB(B), *is reciprocally proportional to the ordinate* BC.

Let AD(D) be another straight line perpendicular to the axis AB(B), and let ordinates CD be drawn to it, so that the abscissae AD along the axis AD(D) are equal to the ordinates BC to the axis AB(B), and the ordinates CD to the axis AD(D) are equal to the abscissae AB along the axis AB(B). Let us call $AD = BC = y$, and $AD = BC = x$; also let $BP = w$ and $B(B) = z$. Then it follows from what I have proved in another place that

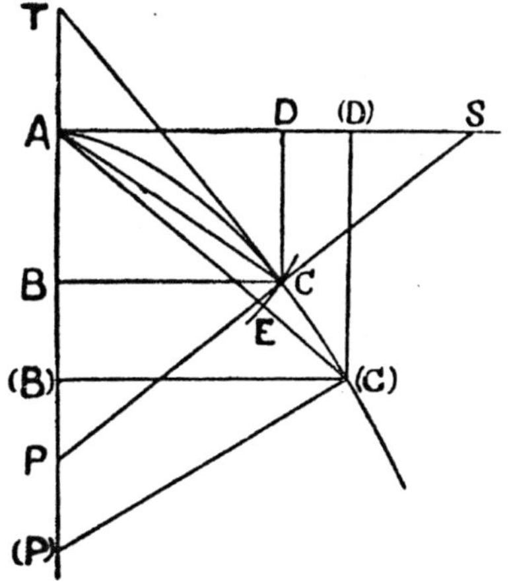

$$\int wz = \frac{y^2}{2}, \text{ or } wz = \frac{y^2}{2d}\ ^{27}$$

But from the quadrature of a triangle it is evident that $\frac{y^2}{2d} = y$; and therefore $wz = y$.

<hr>

[26] See Cantor, III, p. 183; but neither Cantor nor Gerhardt appears to offer any suggestion as to why this date should have been altered.

[27] See foot of next page.

Now, from the hypothesis, $w = b/y$, for thus w and y will be reciprocally proportional to one another.   Hence we have

$$\frac{bz}{y} = y, \text{ and thus } z = \frac{y^2}{b}.$$

But $\int z = x$, hence $x = \int \frac{y^2}{b}$; and from the quadrature of the

parabola $\int \frac{y^2}{b} = \frac{y^3}{3ba}$; hence, $x = \frac{y^3}{3ba}$; and this is the required

equation expressing the relation between the ordinates y and the abscissae x of the curve $C(C)$, which was to be found.   Therefore we consider that the curve has been found and it is analytical; in short, it is the cubical parabola whose vertex is A.

We will therefore see whether the truly remarkable theorem is not true, namely, in the cubical parabola $C(C)$, the intervals BP between the normals to the curve, PC, and the ordinates to the axis, BC, taken along the axis ABP, are reciprocally proportional to the ordinates, BC.

The truth of this is easily shown by the calculus of tangents. For the equation to the cubical parabola is $xc^2 = y^3$; taking c to be the *latus rectum*, and supposing that for c^2 we put $3ba$, or $c = \sqrt{3ba}$, we have $3xba = y^3$.

Now, by Slusius's method of tangents, we have $t = y^3/3ba$, where t is put for BT, the interval along the axis between the tangent and the ordinate.

But $BP = w = \dfrac{y^2}{t}$, and therefore $w = \dfrac{y^2}{\dfrac{y^3}{ba}} = \dfrac{ba}{y}$; hence, the w's

and the y's are reciprocally proportional as was to be proved.

[27] This was obtained in the form $omn.p = y^2/2$, previous to October, 1674, from the Pascal form of the characteristic triangle; it is quoted as a known theorem in the essay dated 29 October, 1675.   See §§ 3, 6.

It is probably at this date that he began to revise his ideas as to d diminishing the dimensions; being forced to reconsider them by the occurrence of such equations as $wz = y$. It is seen in the next paragraph how careful he is to keep his dimensions equal; for he introduces an apparently irrelevant $a(=1)$ for this purpose. It gradually dawns on him that neither $\int$ nor d alters the dimensions, but that a "sum of lines" is really a sum of rectangles, on account of the fact that they are applied in a certain fixed way to an axis; he is not quite certain of this however until well on in the next year, when we find him using $\int dx\, y$.

The artifice of this analysis[28] consisted in obtaining the abscissa from the ordinate; and this idea was never previously thought of. It is not a more difficult question either, if the curve is required in which BP, the interval between the normals and the ordinates, is reciprocally proportional to the abscissae AB.   Indeed, $w = a^2/x$; but $w = y^2/2$; hence, we have

$$y = \sqrt{2 \int w} \text{ or } \sqrt{2 \int \frac{a^2}{x}}.$$

Now $\int w$ cannot be found except by the help of the logarithmic curve.[29]   Hence, the figure that is required is that in which the ordinates are in the subduplicate ratio of the logarithms of the abscissae; and this curve is one of the transcendental curves.

Now, in truth, it is a much harder question,[30] if the curve, in which AP is reciprocally proportional to the ordinate BC is required.

For then $x + w = \dfrac{a^2}{y}$ and $wz = \dfrac{y^2}{2d}$; also $\int z = x$,

or $z = \dfrac{x}{d}$; thus, $w\dfrac{x}{d} = \dfrac{y^2}{2d}$, and $w = \dfrac{y^2}{2d} \cup \dfrac{x}{d}$;

hence, $x + \dfrac{y^2}{2d} \cup \dfrac{x}{d} = \dfrac{a^2}{y}$.

If we suppose that the x's are in arithmetical progression then $x/d = z$ will be constant, and we shall have

$$x + \frac{y^2}{2d} = \frac{a^2}{y} \text{ or } \int x = \int \frac{a^2}{y} - \frac{y^2}{2},$$

therefore

$$\frac{x^2}{2} + \frac{y^2}{2} = \int \frac{a^2}{y} \text{ or } d\,\overline{x^2 + y^2} = \frac{2a^2}{y};$$

[28] It is difficult to see exactly what Leibniz means by this statement; I can only guess at substitution by means of the theorem $wz = y$, the equivalent to the recognition of the fact that $y\,dy/dx.dx = ydy$. The wording is however impersonal, and may mean that he himself had never thought of the idea before.   Barrow has many such theorems for changing the variable.

[29] Required $y = f(x)$, such that $y\,dy/dx = a^2/x$; the solution is $y^2 = 2a^2 \log_e Ax$.   Weissenborn remarks on the omission of the a as being incorrect; from Leibniz's standpoint I cannot agree with him.   Leibniz, from Mercator's work, connects a^2/x with the ordinate of the equilateral hyperbola $xy = a^2$, and its integral with the quadrature of this curve.   The omission of the a^2 only alters the base of the logarithm, and Leibniz merely states that the solution is of a logarithmic nature without attempting to give it exactly.

[30] How does he know until he has tried it?   This rather combats the idea that these were mere exercises; it gives this essay the appearance of being a fair copy intended either for publication or for one of his correspondents.   If this were the case, the errors later in algebraical work are all the more unintelligible.   The idea that Leibniz was a man who was accustomed to writing down his thoughts as he went along does not appeal to me at all; this is the method of the slow-working mind, rather than that of genius.

but, if we join AC, A(C), then these are equal to $\sqrt{x^2+y^2}$; and if with center A and radius AC we describe an arc CE to cut the straight line AE(C) in E, then E(C) will be the difference between AC and A(C); that is, $E(C) = e = d\overline{x^2+y^2}$

$$\therefore e = 2a^2/y.$$

If then it were allowable to assume that the $y's$ were also in arithmetical progression, we should have what was required; yet it seems that it does not make any difference even if the x's have been assumed to be in arithmetical progression. For if we do assume that the x's are in arithmetical progression, it follows that the AD's, or the y's are the reciprocals of the E(C)'s or the e's. Moreover, if they are so at any one time they are so at all times. Also, the sums of an infinite number of reciprocal proportionals, no matter what the progression may be of which they are taken as the reciprocal proportionals; for in this case there is not any consideration of rectangles, where there is need of equal altitudes, but a sum of lines is calculated, that of all the E(C)'s.[31] Hence I see the difficulty arises from the fact that the sum of every e, or every $2a^2/y$, or every E(C), cannot be obtained, unless we know to what progression the y's belong. In this case, that information is not given; for it is necessary that the x's should be in arithmetical progression, and hence that the y's are not so.

On the other hand, if we suppose in the above equation,

$$x + \frac{y^2}{2d} \cup \frac{x}{d} = \frac{a^2}{y},$$

that the y's are in arithmetical progression, then we have

$$x + \frac{y}{dx} = \frac{a^2}{y} \text{ or } xy + \frac{y^2}{dx} = a^2 ;$$

and, finally, by assigning the progression to neither x nor y, we have in general

$$xy + y \,\frac{d\frac{y^2}{2}}{dx} = a^2 \dots\dots\dots\dots\dots\dots\dots (A)$$

But we have not as yet really obtained anything. Let us therefore consider it from the standpoint of "indivisibles"; let PCS produced meet AD in S; then the sum of every AP applied to AB

---

[31] This seems to be the root of the error into which he falls; he has not yet perceived that the e's have to be *applied to some axis*, before he can sum them; and this is to a great extent due to the omission of the dx, taken as constant and equal to unity. He is thus bound to fall back on the algebraical summation of a series.

is equal to the sum of every AS applied to AD ;[32] or calling DS, v, we have

$$dy \int y + dy \int v = dx \int x + dx \int w,$$
$$\text{or } dy \int y + dy \int v = dx \int a^2/y,$$

by the hypothesis of the question.

Now, if we take the y's to be in arithmetical progression, we have

$$\frac{y^2}{2} + \frac{x^2}{2} = dx \text{ Log} y. \quad {}^{33}$$

But just above, making the same supposition that the y's were in arithmetical progression, we had

$$xy + \frac{y^2}{dx} = a^2 \text{ or } dx = \frac{y^2}{a^2 - xy};$$

and now we have

$$dx = \frac{x^2 + y^2}{2 \overline{\text{Log} y}}.$$

Hence at length we obtain an equation, in which x and y alone remain, and unshackled, namely

$$\overline{y^2 + x^2}, \ a^2 - yx = 2y^2 \overline{\text{Log } y};$$

and this equation, since it is determinate, will give the required locus.

This then is an exceedingly remarkable method, for the reason that when it is not in our power to have as many equations as there are unknowns, yet often we shall be able to obtain some more equations, by the help of which we shall be able to eliminate certain terms, as the term dx in this case, which alone stood in our way. Either of the two equations, by itself, contained the whole nature of the locus, although from neither of them could the solution be derived, because so far easy means were lacking; yet the combination of the two equations gave the solution at once.

I see that the same thing could be otherwise obtained by moments; and here there comes to my mind a new consideration that is not altogether inelegant.

[32] From the characteristic triangle, AS: AP $= dx : dy$.

[33] This is of course nonsense. The error seems to arise from the dx being placed outside the integral sign; thus he assumes that dx is constant, while, for the integration, he also assumes that the dy is constant.

We cannot argue from this equation that Leibniz did not at this date appreciate what an infinitesimal was, on account of the infinitesimal being equated to a finite ratio; for since he is assuming that dy is an infinitely small unit, dx really stands for dx/dy.

In the attached figure, let $BC = y$, $FC = dy$; let S be the middle point of FC; **then it is evident that the moment of FC is the** rectangle contained by FC and BS, i. e., the rectangle BFC; this follows from the fact that it is equal to BFC + SFC, and the latter can be neglected as being infinitely small compared to the former.[34]

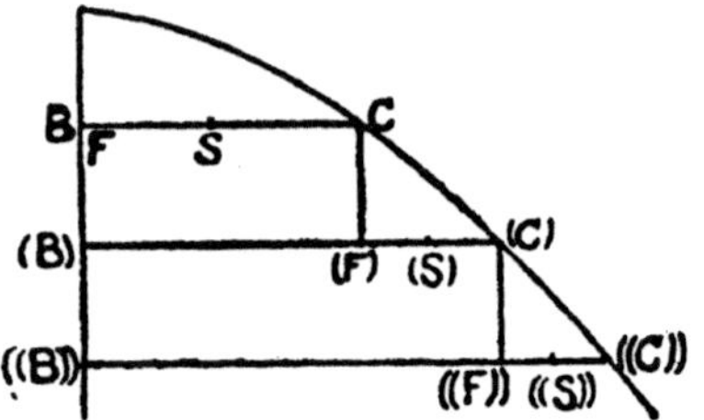

Hence $\int y \, dy = y^2/2$, or the moment of all the differences FC will be equal to the moment of the last term, and $y \, dy = d(y^2/2)$, or $y^2 \, dy = y \, dy^2/2$.

Now, just above, in equation (A), by making x arithmetical, we had

$$y \, d\frac{y^2}{2} = a^2 - xy \,, \text{ or } d\frac{y^2}{2} = \frac{a^2 - xy}{y} \,;$$

but this is the same thing as $y \, dy$; hence $y \, dy = \dfrac{a^2 - xy}{y}$, and therefore

$$\int \overline{y \, dy} = \int \frac{a^2}{y} - \frac{x^2}{2} \,. \quad \text{But we have already found that } \int \overline{y \, dy} = \frac{y^2}{2} ;$$

therefore $y^2 + x^2 = 2 \int \dfrac{a^2}{y}$, as before; i. e., $\overline{dx^2 + y^2} = \dfrac{2a^2}{y}$.

From this there follows something to be noted about these equations, in which occur $\int$ and d, where one quantity, in this case for instance the x, is taken to proceed arithmetically, namely, that we cannot make a change, nor say that the value of x is found, thus, $x = 2(a^2/y) - d \, \overline{y^2}$; for $d \, \overline{y^2}$ cannot be understood unless the nature of the progression of the y's is determinate. But the progression of the y's, in order that it may be used for $d \, \overline{y^2}$, must be such that the x's are in arithmetical progression; hence the dy's depend on the x's, and therefore the x's cannot be found from the dy's. For the rest, by this artifice many excellent theorems with regard to curves that are otherwise intractable will be capable of being investigated, namely, by combining several equations of the same kind.

In order that we may be better trained for really very difficult considerations of this kind, it will be a good thing to attempt just one more, as for instance when the AP's are reciprocally proportional to the AB's.

---

[34] Note the advance in ideas suggested by the words "infinitely small *compared with* the former." Here, of course, the notation BFC is the usual notation of the period for BF.FC, the rectangle contained by BF and FC.

Here $x + w = \dfrac{a^2}{x}$, and $zw = d\dfrac{y^2}{2}$, and $z = dx$; and so we obtain

$$w = \dfrac{d\overline{\dfrac{y^2}{2}}}{z} = \dfrac{d\overline{\dfrac{y^2}{2}}}{dx}, \text{ hence } x + \dfrac{d\overline{\dfrac{y^2}{2}}}{dx} = \dfrac{a^2}{x}.$$

The solution of this is not now difficult; for if we suppose that the x's are arithmetical,[35] we have

$$\int x + \dfrac{y^2}{2} = \int \dfrac{a^2}{x}, \text{ or } x^2 + y^2 = \overline{\mathrm{Log}\, y}. \qquad (36)$$

Hence, $\sqrt{x^2 + y^2} = \mathrm{AC} = \sqrt{2\, \mathrm{Log}\, \mathrm{AD}}$; and this is a simple enough expression for the curve. In this however the AP's are required to be in arithmetical progression; but on the other hand, if the y's are taken to be in arithmetical progression, we have $x + y/dx = a^2/x$; and from this latter the nature of the curve is not easily obtained.

Let us see whether there can be a curve in which AC is always equal to BP; in this case $\sqrt{x^2 + y^2} = w$, and $w = dy^2/2dx$. Let the x's be in arithmetical progression then $(\int \sqrt{x^2 + y^2} =)\ \int \mathrm{AC} = y^2$; this, however, is not sufficient to describe the curve practically, that is to say, by points following one another consecutively. When $x = 1$, let $\mathrm{BC} = (y)$; then $\sqrt{1 + (y^2)} = (y^2)$, or $1 + (y^2) = (y^4)$. Whence (y) may be obtained; thus, from the equation

$$y^4 - y^2 + \tfrac{1}{4} = 1 + \tfrac{1}{4}, \text{ we have } (y^2) = \dfrac{\sqrt{5}}{2}, \text{ or } (y) = \dfrac{\sqrt[4]{5}}{\sqrt{2}}. \qquad (37)$$

Further, in the same way,

$$\underset{\text{AC}}{\sqrt{4 + ((y^2))}} + \underset{\text{A(C)}}{\sqrt{1 + \dfrac{\sqrt{5}}{2}}} = ((y^2));$$

and thus again $((y))$ can be found. By the help of this a third AC can be found, and some sort of polygon can be found, which is more and more like the curve that is required, in proportion as the thing taken for unity is less and less.

[35] Note in general that this is Leibniz's equivalent of the modern phrase, "integrate with respect to x." (For the rest, see fig., p. 93.)

[36] This I think is more likely to be a slip on the part of Leibniz, than a misprint; for in the next line he has AD, which is the correct equivalent of y. Further, AP varies inversely as x, hence the AP's have to be in harmonical progression, not arithmetical, otherwise x is not equal to $x^2/2$. If on the other hand, we assume three errors of transcription, and replace x for y, AB for AD, AB for AP, the whole thing is correct with an arbitrary base.

[37] It is hardly necessary to point out the error in the arithmetical solution of the quadratic; nor is it important. It is however to be noted that if $\mathrm{AC} = v$, the equation reduces to $v^2 = x(x + v)$, and the solution is a pair of straight lines.

That the x's are in arithmetical progression signifies that the motion (in describing it) along the axis AB is uniform. But descriptions that suppose any motion to be uniform are not within our power.[38] For we cannot produce any uniform motion, except a continually interrupted one.

Let us now examine whether $dx\,dy$ is the same thing as $\overline{dxy}$, and whether dx/dy is the same thing as $d\dfrac{x}{y}$; it may be seen that if $y = z^2 + bz$, and $x = cz + d$; then

$$dy = z^2 + 2\beta z + \beta^2, + bz + b\beta, - z^2 - bz,$$

and this becomes $dy = 2z + b\beta$.

In the same way $dx = + c\beta$, and hence

$$dx\,dy = \overline{2z + b}\,c\beta^2.$$

But you get the same thing if you work out $\overline{dxy}$ in a straight-forward manner. For in each of the several factors there is a separate destruction, the one not influencing the other; and it is the same thing in the case of divisors.

Now let us see if there is any distinction when we seek the sums of these things. We have $\int dx = x$, $\int dy = y$, and $\int \overline{dxy} = xy$. If then we have an equation, $dx\,dy = x$ say, then $\int dx\,dy = \int x$. But $\int x = x^2/2$, hence $xy = x^2/2$, or $x/2 = y$; and this satisfies the equation $dx\,dy = x$; for substituting for y its value, $ax\dfrac{dx}{2} = x$, or $a\dfrac{x^2}{2} = x,$ (39) which is known to be true.

In sums these results do not hold good; for $\int x \int y$ is not the same thing as $\int xy$; the reason is that a difference is a single quantity, while a sum is the aggregation of many quantities. The sum of the differences is the latest term obtained. However, from the sums of the factors we can find the sums of products, not indeed as yet analytically, but by a certain method of reasoning; such as Wallis has done in this class of thing, not by proving them, but by a happy method of induction. Nevertheless to find proofs for them would be a matter of great importance.

Suppose $\int \overline{zy}$ to be the sum that is required. Let $\int \overline{zy} = w$, then $zy = \overline{dw}$, and $y = \dfrac{dw}{z}$, and $\int y = \int \dfrac{dw}{z}$. Similarly, $\int z = \int \dfrac{dw}{y}$.

[38] This is strongly reminiscent of Barrow, Lect. I (near the beginning) and Lect. III (near the end).

[39] Leibniz, as a logician, should have known better than to trust a single example as a verification of an affirmative rule.

With regard to infinitesimals note the equation $dx\,dy = x$!

Suppose that $\int y$ is known, $= v$, and that $\int z$ is known, $= \psi$; then y

$= dv = \dfrac{dw}{z}$, and $z = d\psi = \dfrac{dw}{y}$, and $\dfrac{dv}{d\psi} = \dfrac{z}{y}$.   From this it would seem

to follow that $d\dfrac{v}{\psi} = \dfrac{z}{y}$, and therefore that $\dfrac{v}{\psi} = \int \dfrac{z}{y}$.   Therefore $\int \dfrac{z}{y} =$

$\dfrac{\int z}{\int y}$, which is obviously incorrect. [40]   Hence it follows that $\int \dfrac{dv}{d\psi}$

cannot be equal to $\dfrac{v}{\psi}$.

What then can it be? We have to sum the difference for v divided by the difference for ψ. That is, not every one of the differences for, or the whole of, v is to be divided by each single difference for the ψ; this is not so, I say, because each single one of the first set is only divided by the single one of the other set that corresponds to it, and not by all of them. Therefore

$\int \dfrac{dv}{d\psi}$ is not the same as $\dfrac{\int dv}{\int d\psi}$, or $\dfrac{v}{\psi}$.   Will not then $d\dfrac{v}{\psi}$ be something

different from $\dfrac{dv}{d\psi}$? If it is the same, then also $\int d\dfrac{v}{\psi} = \int \dfrac{dv}{d\psi}$, that is

$\dfrac{v}{\psi} = \int \dfrac{dv}{d\psi} = \dfrac{\int dv}{\int d\psi}$, which is absurd.

Similarly, if we can suppose that $\overline{dv\psi} = dv\, d\psi$, then $\int \overline{dv\psi}$, or $v\psi = \int \overline{dv\, d\psi}$.   Now $v\psi = \int dv \int d\psi$; hence, $\int \overline{dv\, d\psi} = \int dv \int d\psi$; which is absurd.

Hence it appears that it is incorrect to say that $dv\, d\psi$ is the same thing as $dv\psi$, or that $\dfrac{dz}{d\psi} = d\dfrac{v}{\psi}$; although just above I stated that this was the case, and it appeared to be proved. This is a difficult point. But now I see how this is to be settled.

If we have v and ψ, and they form some quantity, say $\phi = v\psi$ or v/ψ, and if the values of v and ψ are expressed as rationals in terms of some one thing, for instance, in terms of the abscissa x, then the calculus will always show that the same difference is produced, and that $d\phi$ is the same as $dv\, d\psi$ or $dv/d\psi$. But now I see

<hr>

[40] If Leibniz can see that this equality is "obviously incorrect," what is the use of the argument that has preceded this sentence; for the final result must also be obviously incorrect.

the former can never happen, nor can it come to the latter **by** separation of parts; for example,

$$x+\beta, \frown x+\beta,, -, x, x, \text{ becomes } 2\beta x,$$

which is quite a different thing from

$$x+\beta, -x,, \frown x+\beta, -x \text{ which gives } \beta^2.$$

Hence it must be concluded that $dv\psi$ is not the same as $dv\,d\psi$, and

$$d\frac{v}{\psi} \text{ is not the same as } \frac{dv}{d\psi}. \qquad (41)$$

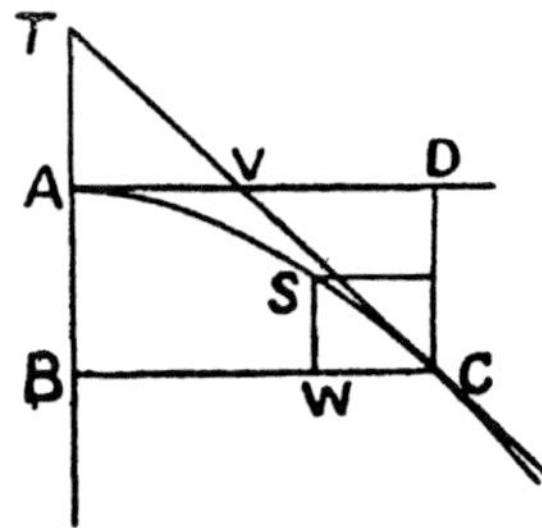

Take an equation of the first degree, $a+bx+cy=0$. Let $DV=\theta$, $AB=x$, $BC=y$, and $TB=t$. Then, by making use of the method of tangents,[42] we have $bt=-cy$, or $t=-cy/b$. In the same way, $\theta=-bx/c$.

Let $WC=w$, and $WS=\beta$, then it is evident that $t/y=\beta/w$, and

$$\text{therefore } w = -\beta\frac{b}{c};$$

$$\text{and in the same way, } \beta = \frac{-wc}{b}.$$

Second degree. $a+bx+cy+dx^2+ey^2+fyx=0$. Making use of the method of tangents, we have

$$bt+2dxt+fyt = -cy-2ey^2-fyx;$$

hence $t=\dfrac{-cy-2ey^2-fyx}{b+2dx+fy}$. From this it is quite evident that t can

[41] Leibniz here justifiably verifies the falsity of his supposition being a general rule by a single breach of it. He uses $v=\psi=x$, and changes x into $x+\beta$; thus,

$$\frac{d(xx)}{dx\ dx} = \frac{(x+\beta)(x+\beta)-xx}{(x+\beta-x)(x+\beta-x)} = \frac{2\beta x}{\beta^2}.$$

Here we see the first idea of the method that is the same as that used by Fermat and, afterward by Newton and Barrow; this consideration, whatever the source, is that which leads him later to the substitution $x+dx$, $y+dy$ in those cases in which Barrow uses a and e.

[42] "ordinando et accommodando," literally setting in order and adapting. It is to be remembered that Sluse gave only a rule, and not a demonstration of the rule. Part of the rule was that, if the equation in two variables contained terms containing both the variables, these terms had to be set down *on each side of the equation.* Thus, for the equation $y^3 = bvv - yvv$ would first of all be written

$$y^3 + yvv = bvv - yvv \ldots\ldots\ldots \text{ ordinando (?)}$$

then each term on the left is multiplied by the exponent of y, and each term **on** the right by that of v, thus,

$$3y^3 + yvv = 2bvv - 2yvv \ldots\ldots\ldots \text{ accommodando (?)}$$

and finally one y on the left, in each term, is changed into a t, where t is the subtangent measured along the y axis.

always be divided by y (and θ by x), and since $w=\beta y/t$, therefore we have

$$w = \frac{\overline{\beta b + 2dx + fy}}{-c-2ey-fx}, \text{ and } y = \frac{-w \; \overline{c+fx,} \frown \overline{\beta b + 2 \; dx}}{f+2e},$$

but from just above $y = \dfrac{-a-bx-dx^2}{c+ey+fx}$, hence we have

$$y = \frac{-w, \overline{c+fx,} \frown \overline{\beta b + 2 \; dx}, \frown \overline{c+fx,,,} + \overline{f+2e} \frown \overline{a+bx+dx^2}}{-w, \overline{c+fx,} \frown \overline{-\beta b + 2dx,} \frown -e} \tag{43}$$

$$= \frac{-w \; \overline{c+fx}, - \overline{\beta b + 2dx}}{f+2e} \, .$$

Hence we have an equation in which there is no longer any y;[44] and all figures that can be formed from this equation by a variation of the letters that stand for the constants can be squared; and also all others that by other methods can be shown to be connected with it.

§ 9.

In the manuscript that follows we must refrain from being critical; for, as suggested by the opening remark, it contains nothing more than random notes, jotted down as they came into Leibniz's mind, as materials for further investigation. In the ten days that have intervened since the date of the last MS., he has either had no spare time for further work on the lines of this last manuscript, or else he has found that he cannot proceed any further usefully until he has perfected the method he had in hand. He therefore reverts to the method of breaking up the

[43] This is hopelessly inaccurate; all except one error, namely, $f+2e$, which should be $\beta f + 2ew$, may be put down to bad transcription. Even if Leibniz's writing were execrable, the correct version of an ambiguous sign (through bad writing) could easily have been settled, *by working through the algebra.* Thus the first of the last pair of values, in Leibnizian symbols should be

$$y = \frac{-w, \overline{c+fx}, -\beta, \overline{b+2dx,,} \frown \overline{c+fx,,,} + \overline{\beta f + 2ew}, \overline{a+bx+dx^2}}{-w, \; c+fx, -\beta, \; b+2dx,, \; \smile -e,}$$

with a similar correction in the second value.

[44] Even if Leibniz had worked out the correct result, and obtained what he was trying for, namely, w/β in terms of x, he would have got a very lengthy quadratic, and the roots would be quite beyond his power to use at any time. But he convinces himself that he can thus find the quadrature of any conic, or figures that can be reduced to them.

figure into triangles by means of a set of lines meeting in a point, coupled with the ideas of the moment and the center of gravity, in order to try to obtain further general theorems for *analytical use*. In this way, he again comes across the differentiation of a product in the form of an "integration by parts"; but he does not recognize in it the differentiation of a product, for he says that as he has obtained this before he can get nothing new from it. He is still wasting his energies over the idea of obtaining dy/dx as an explicit function of x, for the purposes of *integration* or quadratures. The fact that he can use the method of Slusius *as an unproved rule* seems to have hidden from him the necessity of pushing on his investigations with regard to the laws of *differentiation*, or the direct tangent method.

21 November 1675.

>*Pro methodo tangentium inversa et aliis tetragonisticis specimina et inventa. Trigonometria indivisibilium. Aequationes inadaequatae. ordinatae convergentes. Usus singularis Centri gravitatis.*

>[Examples and discoveries by means of the inverse method of tangents and other quadratures. Trigonometry of indivisibles. Inadequate equations. Converging ordinates. Special use of the Center of Gravity.]

Subject-matter for a new consideration of the Center of Gravity method, as follows:

A segment AECD having been broken up into infinite triangles, AEC, ACF, etc., let the center of gravity of each of these triangles be found; this is a simple matter, for the center of gravity is always distant from the base a third of the altitude. Then, since the path of the center of gravity multiplied by the area of the triangle is equal to the solid formed by its rotation, and also since the products of the AH's and the infinitesimal parts of the axis are twice the areas of the triangle, also it is plain that the AG's multiplied by the distances of the centers of gravity of the triangles AEC from the axis are equal to the moment of the segment about the axis; by the help of this idea a number of things can be at once obtained in two ways: first, by taking some general figure and making a general calculation, and then so expressing it that the center of gravity can be easily found; in this way we may obtain the moments of spaces which would be a matter of difficulty otherwise, if they were investigated by the ordinary method of ordinates. Secondly, on the other hand, if figures of which the moments are easily obtained in the ordinary way are treated by this method, we shall arrive at certain very difficult curves, the dimensions of which can always be deduced from some that are easier. Here then we have a remarkable rule, by the help of which useful properties can always be obtained from any method however complicated. It is often useful when problems arise that we know are naturally simple, and from other reasons are soluble; for thus many notable cases are discovered. See what Tschirnhaus noted about the Hastarian line.

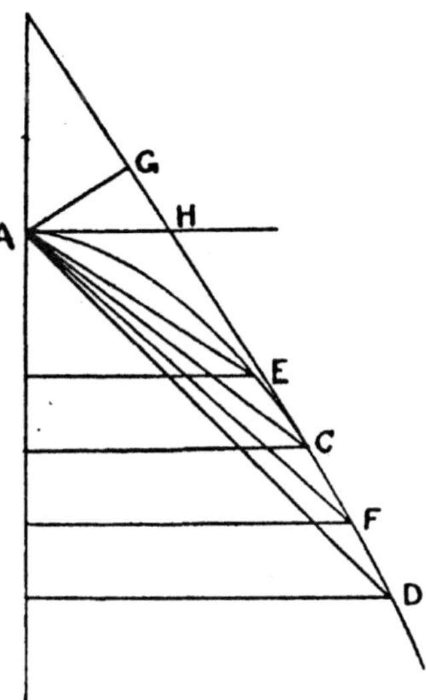

In irregular problems, such as cannot be treated in a straightforward manner or reduced to an equation that is sufficiently determinate, because, say, something has to be done inversely, it is useful to compare several ways with one another, of which the results should be identical. This seems to be useful for the inverse tangent method. Here is a case in point.

The figure, in which BP and AT are reciprocally proportional, is required.

Let $TB = t$, then $AT = t - x$,
and $BP = a^2/(t - x)$.

If this is multiplied by t, we have

$$\square TBP = ta^2/(t - x) = a^2 + a^2x/(t - x) = y^2$$

hence, $ta^2 = ty^2 - xy^2,$

or $t = xy^2/(a^2 - y^2)$;[45] and therefore $t/x = y^2/(a^2 - y^2)$, or all the t's together equal the moment about the vertex of every $y^2/(a^2 - y^2)$.

[45] There is a mistake in sign; $a^2 - y^2$ should be $y^2 - a^2$; hence the work that follows is also wrong.

But from other reasons, all the TP's applied to the axis are equal to the TC's applied to the curve.

Now $t/y = \beta/w$, and therefore $w = \dfrac{\beta y}{t = \dfrac{y^2 x}{a^2 - y^2}} = \dfrac{\overline{\beta a^2 - y^2}}{xy}$.

But $\int w = y$, therefore

$$\int \frac{\overline{\beta a^2 - y^2}}{xy} = y \dots\dots\dots\dots\dots\dots\dots (A)$$

Further, $wx = \dfrac{\overline{\beta a^2 - y^2}}{y}$, and $\int wx = yx - \int y\beta$,

hence, $$\int \frac{\overline{\beta a^2 - y^2}}{y} = yx - \int y\beta \dots\dots\dots\dots\dots\dots (B)$$

Also $w = dy$, $dy = \dfrac{\overline{\beta a^2 - y^2}}{xy}$, and therefore

$$xy = \frac{\overline{\beta a^2 - y^2}}{dy = w} = \int \overline{y}\beta + \int \frac{\overline{\beta a^2 - y^2}}{y} .$$

Now if we suppose that the y's are in arithmetical progression, then $w = dy$ is constant and β is variable;

hence, $$\beta = \frac{\displaystyle\int \overline{y\beta + \beta\dfrac{a^2 - y^2}{y}}}{a^2 - y^2}, \quad d\,\overline{\beta a^2 - y^2} = \frac{a^2 \beta}{y} .$$

But from equation (B), $\beta\dfrac{a^2 - y^2}{y} + \beta y = \overline{dyx}$

hence, $$\beta\frac{a^2}{y} = \overline{dyx}.$$

We have thus obtained two equations that are mutually independent, the first

$$\frac{dx}{dy} = \frac{yx}{a + y,\, a - y} \overset{(46)}{} \dots\dots\dots\dots\dots (1)$$

and the second

$$\overline{dyx} = \frac{dx\,a^2}{y} . \dots\dots\dots\dots\dots\dots\dots (2)$$

[46] Although the variables are separable, Leibniz does not recognize the fact that he can make use of this. For later he states that the solution of a problem cannot be obtained from a single equation. In this case we have

$$\frac{dx}{x} = \frac{y\;dy}{y^2 - a^2} = \frac{dv}{v}, \text{ if } y^2 - a^2 = \pm\, v^2.$$

Supposing this substitution to have been effected, Leibniz would have concluded that $x = v$, and would have stated that he had solved the problem.

But here again he has made an unfortunate choice, for the origin (A) cannot fall on any of the curves $Cx = v$ or $Cx^2 \pm y^2 = \pm\, a^2$, which is the general solution of the equation. Hence the problem is impossible.

Let us seek to obtain others in addition, such as

$$\int t \, dy = \int y \, dx.$$

Now this furnishes us with nothing new; but $\int tw + \int xw = xy$

or $t \, dy + x \, dy = \overline{dxy}$, and $t = \dfrac{dx}{dy} \, y$; hence the latter $= \dfrac{\overline{dxy - x \, dy}}{dy}$.

Therefore $\overline{dx} \, y = \overline{dxy} - x \, \overline{dy}$.

Now this is a really noteworthy theorem and a general one for all curves. But nothing new can be deduced from it, because we had already obtained it.

However, from another principle we shall obtain a new theorem; for it is known that the sum of every $BP = BC^2/2$; that is to say, $BP = \dfrac{a^2}{t - x}$, $t = \dfrac{\beta y}{w} = \dfrac{dx}{dy} \, y$, and therefore

$$BP = \frac{a^2 \, dy}{\overline{dx} \, y - \overline{dy} \, x} = \frac{\overline{dy^2}}{2} \, \ldots\ldots\ldots\ldots\ldots\ldots (3)$$

We therefore have two equations, in which dx occurs, namely, the first and the third; by the help of these, by eliminating dx, we shall have an equation in which only one of the unknowns remains shackled; thus from equation (1), we have $dx = \dfrac{\overline{dy} \, yx}{a^2 - y^2}$, and now from equation (3), we get $\overline{dx} \, y \, \overline{dy^2} - dy \, \overline{dy^2} x = 2a^2 \, dy$. Hence,

$$dx = \frac{2a^2 \overline{dy} + dy \, \overline{dy^2} x}{y \, \overline{dy^2}}.$$

We have therefore an equation between the two values of dx, in which only the y remains shackled. From this, by assuming the y's to be in arithmetical progression, that is that $dy = \beta$ a constant, and $\overline{dy^2} = z$, and $z = z^2/2 = y^2$; $z = \sqrt{2} \, y = \overline{dy^2}$.[47] Thus we have obtained what was required.

We have here a most elegant example of the way in which problems on the inverse method of tangents are solved, or rather are reduced to quadratures. That is to say that the result is obtained by combining, if possible, several different equations, so as to leave one only of the unknowns in the tetragonistic shackle. This can be done by summing ordinates in various ways, or on the other hand, instead of ordinates, converging or other lines.

[47] This is quite unintelligible to me as it stands; query, is it an accurate transcription?

NOTE. If, instead of x or y, some other straight line can be found, either one that is oblique, or one of a number converging to the same point, by the employment of which one only of the unknowns is left in bonds, it may be employed with safety. Take for instance the case of finding the relation for the AP's; here the sum of AP's applied to the axis is half the square on AC. Whenever the formula for the one unknown that is left in shackles is such that the unknown is not contained in an irrational form or as a denominator,[48] the problems can always be solved completely; for it may be reduced to a quadrature, which we are able to work out; the same thing happens in the case of simple irrationals or denominators. But in complex cases, it may happen that we obtain a quadrature that we are unable to do. Yet, whatever it may come to, when we have reduced the problem to a quadrature, it is always possible to describe the curve by a geometrical motion; and this is perfectly within our power, and does not depend on the curve in question. Further, this method will exhibit the mutual dependence of quadratures upon one another, and will smooth the way to the method of solving quadratures. Meanwhile I confess that it may happen that there may be need for a very great number of inadequate equations (for so I call them, when there is need for many to solve the problem, although each alone would suffice provided it could be worked out by itself), in order to completely free one of the unknowns from its shackles. For, unfortunately, a solution cannot be obtained from a single equation, unless one of the terms is free from shackles; and if this term appears oftener, then not unless it is freed at least once. Thus there may be a great number of inadequate equations to be found; and we have to examine which of them are in some way independent of the others, i. e., such as cannot be derived from one another by a simple manipulation; for instance, the sum of all the AP's and the sum of all the AE's.

A new kind of Trigonometry of indivisibles, by the help of ordinates that are not parallel but converge.

Let B be a fixed point; let BDC be a very narrow triangle standing upon a curve; let DE be the perpendicular to BC; from the point

---

[48] This is tantamount to a confession by Leibniz that he cannot explicitly integrate $\int a^2/y$, although he knows that it is logarithmic or reduces to the area under the hyperbola; for he has given this in the MS. for Nov. 11.

B let BA, perpendicular to BC or parallel to DE, be drawn to meet the tangent AHDC, and let BH be the perpendicular to the tangent DC produced.

Then the triangles CED, CHB, BHA are similar; hence we have

$$BH/CE = HA/DE = BA/CD,$$

and therefore BH, DE = CE, HA,
            and BH, CD = CE, BH.

Hence it follows that the sum of the triangles or the area of the figure is equal to the products of the AB's into the CE's, or the differences of the ED's, and lastly

AH, CD = DE, BH.[49]

Further, CH/CE = HB/DE = CB/CD;
hence, again, CH, DE = CE, HB, and HB, CD = DE, CB; i. e., the area of the triangle, as is in itself evident, is equal to itself. Lastly, CH, CD = CE, CB; and this last result seems to be worth noting for the case of a Trochoid.

For, if by the rolling of a curve DC on a fixed plane CA, a trochoid curve is described by the point B fixed in DC, and it is given that the ordinate of the trochoid drawn to the fixed plane CA is BH, then the sum of the intercepts CH applied to DC will be equal to the sum of the CB's applied to their own differences. Now if any ordinates are applied to their own differences, the same thing is always produced as in the case where we try to find the moment of the differences about the axis, which is the same as the moment when we take the sum of each, or the maximum ordinate, into the distance of its center of gravity from the axis, i. e., its middle point, that is to say into half itself. Finally this is equal to half the square on the maximum ordinate. Therefore we can always obtain the sum of all the rectangles BC, CE, which is always equal to half the square on BC, or to the sum of all the BP's applied to the axis in F, where CP is the normal to the curve DC.

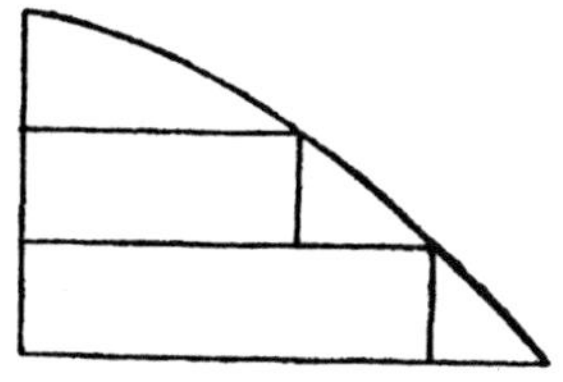

----

[49] There are several errors in the letters in this paragraph, which are probably due to transcription; thus, an E for a (? badly written) B, an H for an A, etc., would be quite an easily-imagined error, provided the work was not verified during transcription.

§ 10.

Leibniz now directs his attention to the direct method of tangents, and proceeds to generalize the methods of Descartes. Is it only a coincidence that Barrow uses this method regularly, the curve that he is especially partial to being the rectangular hyperbola? Weissenborn suggests the same coincidence occurs with respect to the method of Newton, who uses analytical approximations; but if there is anything in either of these suggestions. I think that the Barrovian idea, which is purely for the construction of tangents, is much nearer to that of Leibniz in this manuscript than is the Newtonian.

However this may be, Leibniz is at last beginning to consider the point as to the method by which the principle of Sluse is obtained. He ascribes it to a development of the method of Descartes; but in this connection I cannot get out of my head the suggestion raised by Barrow's use of the first person *plural,* "frequently used by *us,*" in the midst of a passage that is written, contrary to his usual custom, in the first person *singular* throughout, where he describes the differential triangle and the "*a* and *e*" method. I consider that Sluse has enunciated a *working rule* for tangents, which he has generalized by observation of the results obtained by the use of the "*a* and *e*" method; and that this method had been circulated by Barrow some time before the publication of the *Lectiones Geometricae,* although I confess that I have not found any record of this, nor any distinct evidence of a correspondence between Barrow and Sluse; but there is more than a suggestion of this in the fact that Sluse's article was published in the *Phil. Trans.* for 1672.

It seems more than strange to me that there should be such a prolific crop of differential calculus methods within a couple of years of the work of Barrow in all sorts of places, raised by many different people, and that none of them alludes to the general seed-merchant, as I consider Barrow to have been.

22 Nov. 1675.

> *Methodi tangentium directae compendium calculi, dum jam inventis aliarum curvarum tangentibus utimur. Quaedam et de inversa methodo.*

> [Compendium of the calculus of the direct method of tangents, together with its use for finding tangents to other curves. Also some observations on the inverse method.]

In that which I wrote on Nov. 21, I noted down those things which came to my mind concerning the method of tangents. Returning to the subject, let ACCR and QCCS be two curves that cut one another in one, two, or more points C,C; let AB(B) be the axis; let $AB=x$ be the ordinates, and $BC=y$ the abscissae; then we shall have two equations to the two lines, each in terms of these two principal unknowns. Now if these two equations have equal roots, or the equations have equal values, then the lines will touch one another. Instead of the line QCCS, Descartes chooses the arc of a circle VCCD, whose center is P, so that PC is the least of all the lines that can be drawn from the point P. It will come to the same thing, and often more simply, if we take not the arc of a circle but the tangent line TC(C), that is the greatest of all those that can be drawn from a given point T to the curve.

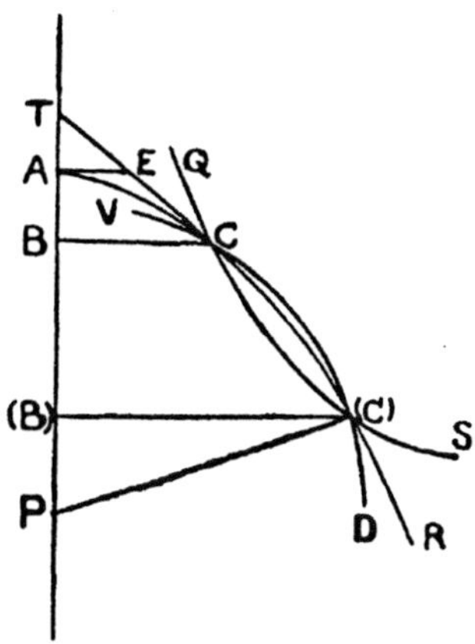

Let $TA=b$, $AE=e$, be assumed as given; required to find AB, BC. The two equations are, the one for the curve AC(C), namely,

$$ax^2 + cy^2 + \text{etc.} = 0,$$

and the other to the straight line TC(C) which, on account of the relation $TA/AE = TB/BC$, will be $b/e = (b \pm x)/y$

$$\text{or } \pm x = (b/e)y - b \quad \text{or } y = \pm (e/b)x + e.$$

Thus the value of either one or other of the unknowns can always be obtained explicitly, and thus can be worked out immediately without raising the degree of the equation of the given curve AC(C); and then at once we shall obtain an equation giving the unknown that alone remains, so that we may determine the condition for equal roots. Doubtless this is the principle of Sluse's method.

If however the arc of the circle whose center is P is used, following Descartes, then the new equation, for the circle, will be as follows: let the radius $PC = s$, and $PB = v - x$, and we have $s^2 = y^2 + v^2 + x^2 - 2vx$. Hence it is clear that we have the choice of either a circle or a straight line; and when, in the equation to the given curve, only an even power of y appears (as can always be made to happen in the case of the conics), then it will be more convenient to use equations to circles; for thus, by the help of the two values of y^2, the unknown x can be immediately worked out; but, in general for all equations to curves expressed by a rational relation, the method of the straight line may be usefully employed.

Hence I go on to say that not only can a straight line or a circle, but any curve you please, chosen at random, be taken, so long as the method for drawing tangents to the assumed curve is known; for thus, by the help of it, the equations for the tangents to the given curve can be found. The employment of this method will yield elegant geometrical results that are remarkable for the manner in which long calculation is either avoided or shortened, and also the demonstrations and constructions. For in this way we proceed from easy curves to more difficult cases, and an equation to a curve being supposed known, it is always possible to choose an equation to some other curve whose tangents are known, by the help of which one of the unknowns can be worked out very easily.

Thus, if it is given that $hy^2 + y^3 = cx^3 + dx^2 + ex + f$ is the equation to a curve of which the tangents are required, assume a curve of which the equation is $hy^2 + y^3 = gx + q$, for that of which the tangents are already known; eliminating y, we have an equation such as $gx + q = cx^3 + dx^2 + ex + f$. This can be determined for two equal roots, either by Descartes's method of comparisons, or Hudde's by means of an arithmetical progression; and thus by working out the value of x, the value of either g or q may be found; and one of the two letters q or g can be chosen arbitrarily.[50] Hence, a way of describing that other curve that touches the given curve is obtained; now, when this is described, let the tangent be drawn at the point which is common to it and the proposed curve, which tangents we

[50] The method of Hudde appears to be similar in principle to that of Sluse, while that of Descartes was the construction of the derived function by assum-

have supposed to be already known; then this tangent will touch the given curve.

I think that, in general, the calculation will be possible by this method of assuming a second curve, as we have done in this case, which evidently works out one of the unknowns. Hence I fully believe that we shall derive an elegant calculus for a new rule of tangents, which in addition may be better than that of Sluse, in that it evidently works out immediately one of the two unknowns, a thing that the method of Sluse did not do. Now this very general and extensive power of assuming any curve at will makes it possible, I am almost sure, to reduce any problem to the inverse method of tangents or to quadratures. Indeed let any property of the tangents to a curve be given, and let the relation between the ordinates and the abscissae be required. Then an equation can be derived, which will contain the principal unknowns, x, y, and always two others as incidentals, such as s and v, or b and e, or the like; now, as the equation contains the property of the tangents, by which s and b may be expressed so as to have a relation to the tangents, assume in this case any new curve chosen arbitrarily, and then s and v will also have some known relation to this curve. By means of the equation to the arbitrarily chosen curve, we shall be able to replace the given property of tangents in favor of the curve required, namely, by removing one or other of the unknowns; and by thus reducing the problem to such a state the inverse calculation will come out the more easily.

The whole thing, then, comes to this; that, being given the property of the tangents of any figure, we examine the relations which these tangents have to some other figure that is assumed as given, and thus the ordinates or the tangents to it are known. The method will also serve for quadratures of figures, deducing them one from another; but there is need of an example to make things of this sort more evident; for indeed it is a matter of most subtle intricacy.

ing roots, forming the sum of the quotients of the function divided by each of the assumed root-factors in turn, and comparison with the original function. Both therefore reduce to finding the common measure of the equation to the curve (where the right-hand side is zero) and the differential of it.

Leibniz, however, strange to say, does not note that by taking one of his arbitrary constants, q, equal to f, the equation has its degree lowered in the particular case he has chosen.

The manuscripts mentioned above seem to be all that were found by Gerhardt belonging to the period 1673-5. I feel that it is a great pity that they were not given in full, or at least a little more fully. For instance, Gerhardt mentions that Leibniz in the MS. of August 1673 constructs the so-called characteristic triangle, but does not give Leibniz's figure in connection. This figure should have been given; for the figure given in October 1674 is not the characteristic triangle as given by Leibniz in the "postscript" (§ 1), or the *Historia* (§ 2), but it *is the Pascal diagram* (assuming that the figure given by Cantor is the correct one). It would be useful to know the date at which Leibniz drops the Pascal diagram in favor of one or other of the Barrow diagrams.

It is to be noticed at this date that Leibniz uses one infinitesimal only, and *verifies* that the method of Descartes comes out correctly in the simple case of the parabola; but he is not satisfied with the generality of the method of neglecting the vanishing quantities.

Again, the second manuscript of October 1674 appears to be immensely important; especially as it contains the groundwork of some of the later manuscripts. Judging by the little that is given of it, it would seem to be most desirable that fuller extracts, at least, should have been given. It is a matter for remark that this manuscript is a long essay on series. Can this possibly have had anything to do with the fact that it is not given in full?

V.

MANUSCRIPTS OF THE PERIOD 1676, 1677, AND A LATER UNDATED MANUSCRIPT.

§§ 11-15.

Between the date of the manuscript last considered and the one which follows there is a gap of seven months, for which Gerhardt does not appear to have found anything. This is very unfortunate; for in this interval Leibniz has attained to the important conclusion that *the true general method of tangents is by means of differences.* We saw that in November 1675 he had *started* to investigate more thoroughly the direct method of tangents; but the method is that of the auxiliary curve, and there is no indication whatever of the characteristic triangle. Does this interval correspond with the time taken by Leibniz for his final reading of Barrow from Lect. VI to Lect. X, comparing all the geometrical theorems with his own notation? Or is it only a strange coincidence that Leibniz's order is the same as that of Barrow, first the auxiliary curve, and lastly the method of differences? One could form a more definite opinion, if Leibniz had given a diagram for the first problem he considers, the one in the next following manuscript, which amounts to the differentiation of an inverse sine. Such a diagram he must have had beside him as he wrote; for I think the reader will find that he wants one to follow the argument; with the idea of verifying this argument, I have not endeavored to supply the omission.

The consideration of the direct method of tangents is apparently, however, only as a means and not as an end; for Leibniz harks back to the inverse method, and to the catalogue of quadrable curves, which he seems to say he has in hand.   It is not until November 1676 that he seems to be coming into his own; and it is not until July 1677 that he has a really definite statement of his rules.   On the other hand, in July 1676, he is consistently using the differential factor with all his integrals, and before the end of that year he has the differential of a product, whether obtained as the inverse of his theorem $\int y\, dx = xy - \int x\, dy$, or by the use of the substitution $x + dx,\ y + dy$, is not certain; but this substitution appears in the manuscript for November 1676.   Finally, in July 1677, appears the general idea of the substitution of other letters, in order to eliminate the difficulty caused by the appearance of the variable under a root sign or in the denominator of a fraction; and with this the whole thing is now fairly complete for all *algebraical functions*.   There is as yet no equally clear method for the treatment of exponentials, logarithms, or trigonometrical functions; for the latter he refers to a geometrical diagram, strongly reminiscent of Barrow.

§ 11.

26 June, 1676.

Nova methodus Tangentium.

[New method of tangents.]

I have many beautiful theorems *with regard to the method of tangents both direct as well as inverse*. Descartes's method of tangents depends on finding two equal roots, and it cannot be employed, except in the case when all the undetermined quantities occurring in the work are expressible in terms of one, for instance, in terms of the abscissa.

But the true general method of tangents is by means of differences. That is to say, the difference of the ordinates, whether direct or converging, is required. It follows that quantities that are not amenable to any other kind of calculus are amenable to the calculus of tangents, so long as their differences are known. Thus if we are given an equation in three unknowns, in which x is an abscissa, y an ordinate, and z the arc of a circle of which x is the sine of the complement, e. g., the equation $b^2 y = cx^2 + fz^2$. To find the next consecutive y, in place of x take $x + \beta$, and in place of z take $z - dz$, or, since $\overline{dz} = \dfrac{\beta r}{\sqrt{r^2 - x^2}}$, we may take $z - \dfrac{\beta r}{\sqrt{r^2 - x^2}}$; [51]

hence we have

$$b^2 (y) = cx^2 + 2cx\beta + c\beta^2 + fz^2 - \frac{2fz\beta r}{\sqrt{r^2 - x^2}} + \frac{\beta^2 r^2}{r^2 - x^2}$$

Hence the difference between y and (y) is given by

$$\pm b^2 y \mp b^2 (y) = + 2cx\beta - \frac{2fz\beta r}{\sqrt{r^2 - x^2}} = b^2 \,\overline{dy} ;$$

Therefore $\quad \dfrac{dy}{\beta} = \dfrac{\mp 2cx\sqrt{r^2 - x^2} \mp 2fzr}{b^2 \sqrt{r^2 - x^2}} = \dfrac{t}{y} = \dfrac{tb^2}{cx^2 + fz^2}.$

From this the flexure or sinuosity of the curve can be found, according as now $2cz\sqrt{r^2 - x^2}$, now $2fzr$ predominates; for when they are equal, the ordinate on that side on which it was previously the greater then becomes the less. It is just the same, if several other undetermined quantities, such as logarithms and other things occur, no matter how they are affected, as for instance in the equation $b^2 y = cx^2 + fz^2 + xzl$, where z is supposed to be an arc, and l a logarithm, x the sine of the complement of the arc, and y the number of the logarithm, b being the radius and unity, equal to r. Also it is just the same, whenever an undetermined transcendental has been derived from some dimension or quadrature that has not been investigated.[52]

For the rest, many noteworthy and useful theorems now arise from the foregoing by the inverse method of tangents. Thus general equations, or equations of any indefinite degree may be formed, at first indeed in two unknowns, x and y, only. But if in this way the matter does not work out satisfactorily, it will easily do so when

[51] In this and the following line I have corrected two obvious misprints; they are evidently not the fault of Leibniz, for the lines that follow from them are correct.

[52] There is some doubt here as to whether Leibniz could have given an example; but it must be remembered that these are practically only notes, mostly for future consideration.

the tables which I am investigating are finished; then it will be possible to take one or more other letters, and to take the difference as an arbitrary known formula, and when this is done it is certain that finally in any case a formula will be found such as is required, and in this way also a curve which will satisfy the conditions given; but in truth the description of the curve will need diagrams for these symbols, representing the sums of the arbitrarily chosen differences.

Now once a curve is found having the tangent property that we want, it will be more easy afterwards to find simpler constructions for it. We have this also as a convenient means enabling us to use many quantities that are transcendent, yet depending the one on the other, such for example as are all those that depend on the quadrature of the circle or the hyperbola. From these investigations it will also appear whether or no other quadratures can be reduced to the quadrature of the circle or the hyperbola. Lastly, since the finding of maxima and minima is useful for the inscription and circumscription of polygons, hence also, by employing these transcendent magnitudes, convergent series can be found, and in the same way their terminations; or of any quantities formed in the same way. However in that case it may not be so easy to argue about impossibility; at least indeed by the same method. Only I do not see how we can find whether from the quadrature of the circle, say, any sum can be found, when no quantity depending on the dimensions of the circle enters into the calculation.

§ 12.

July, 1676.

Methodus Tangentium inversa.
[Inverse method of tangents.]

In the third volume of the correspondence of Descartes, I see that he believed that Fermat's method of Maxima and Minima is not universal; for he thinks (page 362, letter 63) that it will not serve to find the tangent to a curve, of which the property is that the lines drawn from any point on it to four given points are together equal to a given straight line.

[Thus far in Latin; Leibniz then proceeds in French.]

Mons. des Cartes (letter 73, part 3, p. 409) to Mons. de Beaune.

"I do not believe that it is in general possible to find the converse to my rule of tangents, nor of that which Mons. Fermat uses, although in many cases the application of his is more easy than mine; but one may deduce from it *a posteriori* theorems that apply to all curved lines that are expressed by an equation, in which one of the quantities, *x* or *y*, has no more than two dimensions, even if the other had a thousand. There is indeed another method that is more general and *a priori*, namely, by the intersection of two tangents, which should always intersect between the two points at which they touch the curve, as near one another as you can imagine; for in considering what the curve ought to be, in order that this intersection may occur between the two points, and not on this side or on that, the construction for it may be found. But there are so many different ways, and I have practised them so little, that I should not know how to give a fair account of them."

Mons. des Cartes speaks with a little too much presumption about posterity; he says (page 449, letter 77) that his rule for resolving in general all problems on solids has been without comparison the most difficult to find of all things which have been discovered in geometry up to the present, and one which will possibly remain so after centuries, "unless I take upon myself the trouble of finding others" (as if several centuries would not be capable of producing a man able to do something that would be of greater moment).

(Page 459.) The question of the four spheres is one that is easy to investigate for a man who knows the calculus. It is due to Descartes, but as it is given in the book, it appears to be very prolix.

The problem on the inverse method of tangents, which Mons. des Cartes says he has solved (Vol. 3, letter 79, p. 460):

[Leibniz then continues in Latin]

EAD is an angle of 45 degrees. ABO is a curve, BL a tangent to it; and BC, the ordinate, is to CL as N is to BJ. Then

$$CL = \frac{BC = ny}{BJ = y - x}, \quad CL = t,$$

hence,
$$t = \frac{ny}{y - x}, \quad \frac{n}{t} = \frac{y - x}{y} = 1 - \frac{x}{y}$$

hence,
$$\frac{x}{y} = \frac{t - n}{t}; \text{ but } \frac{t}{y} = \frac{\overline{dx}}{dy};$$

therefore
$$\frac{\overline{dx}}{dy} = \frac{n}{y - x}, \text{ or } \overline{dx}\, y - x\, \overline{dx} = \overline{dy}\, n;$$

hence
$$\int \overline{dx}\, y - \int x\, \overline{dx} = -n \int dy.$$

Now, $\int \overline{dy} = y$, and $\int \overline{x\,dx} = x^2/2$, and $\int \overline{dx\,y}$ is equal to the area ACBA, and the curve is sought in which the area ACBA is equal to $(x^2/2) + ny = (AC^2/2) + nBC.$[53]

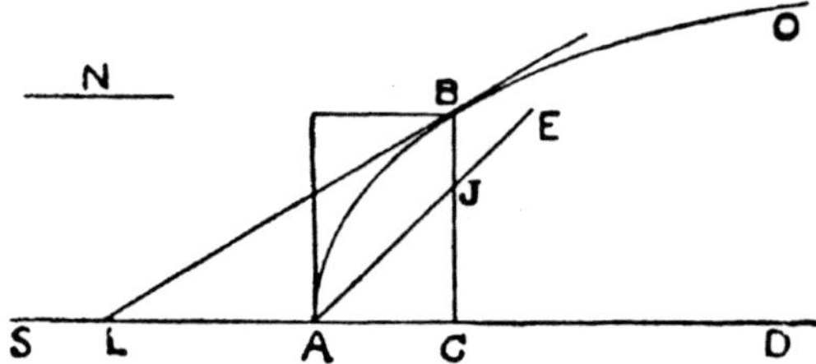

Let this $x^2/2$, i. e., the triangle ACJ be cut off from the area, then the remainder AJBA should be equal to the rectangle ny.

The line that de Beaune proposed to Descartes for investigation reduces to this, that if BC is an asymptote to the curve, BA the axis, A the vertex, AB, BC, fixed lines, for BAC is a right angle.

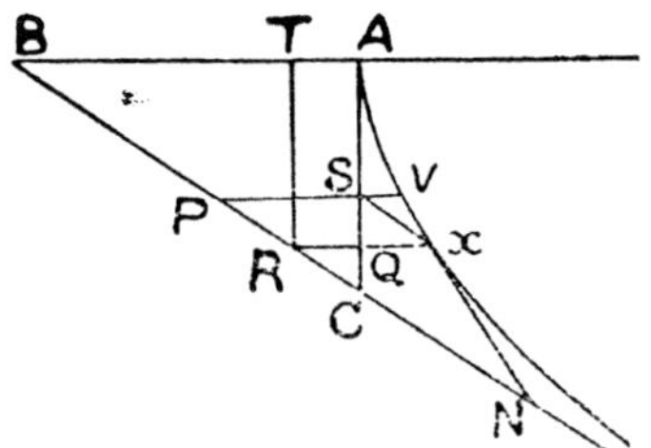

Let RX be an ordinate, XN a tangent, then RN is always to be constant and equal to BC; required the nature of the curve.

This is how I think it should be done.

Let PV be another ordinate, differing from the other one RX by a straight line VS, found by drawing XS parallel to RN; then

[53] Leibniz has a footnote to this manuscript: "I solved in one day two problems on the inverse methods of tangents, one of which Descartes alone solved, and the other even he owned that he was unable to do."

This problem is one of them, the first mentioned in the footnote given by Leibniz. But it requires a stretch of imagination to consider Leibniz's result as a solution. For he ends up with a geometrical construction, that is at least as hard as the construction that can be made by the use of the original data. There is of course the usual misprint that one is becoming accustomed to; but there is also the unusual, for Leibniz, mistake of using his data incorrectly. Starting with the hypothesis that $BC : CL = N : BJ$, he writes $CL = N.BC/BJ$ (correcting the omission of the factor N), instead of $CL = BC.BJ/N$.

The solution of the problem is $y + n \log(y - x + n) = 0$, as originally stated, or $x = n \log(n - y + x)$, if we continue from Leibniz's erroneous result $dx/dy = n/(y - x)$.

The point to be noted, however, is that Leibniz does not remark that "this curve appertains to a logarithm."

the triangles SVX, RXN are similar, RN $=t=c$, a constant, RX $=y$, SY $=dy$, and therefore

$$\frac{\overline{dy}}{dx} = \frac{y}{t=c}\;;\;\text{hence } cy = \int \overline{y\,dx}\;\;\text{ or } c\,\overline{dy}=y\,\overline{dx}. \qquad \text{\footnotesize 54}$$

If AQ or TR $=z$, and AC $=f$, while BC $=a$;

then, $\dfrac{\text{AC}}{\text{BC}} = \dfrac{f}{a} = \dfrac{\text{TR}}{\text{BR}} = \dfrac{z}{x}\;;\;$ and thus $x = \dfrac{az}{f}$.

If $\overline{dx}$ is constant, then $\overline{dz}$ is also constant. Hence

$c\,dy = \dfrac{a}{f}y\,\overline{dz}$, or $cy = \dfrac{a}{f}\int \overline{y\,dz}$, and $cy\,\overline{dy} = \dfrac{a}{f}y^2\,\overline{dz}$, therefore

$c\dfrac{y^2}{2} = \dfrac{a}{f}\int y^2\,\overline{dz}$. Hence we have both the area of the figure and the

moment to a certain extent (for something must be added on account of the obliquity) ; also

$cz\,\overline{dy} = \dfrac{a}{f}yz\,\overline{dz}$, and therefore $c\int z\,\overline{dy} = \dfrac{a}{f}\int yz\,\overline{dz}$.

Also $\dfrac{c\overline{dy}}{y} = \dfrac{a}{f}dz$, and hence, $c\int \dfrac{\overline{dy}}{y} = \dfrac{a}{f}z$. Now, unless I am

greatly mistaken, $\int \dfrac{\overline{dy}}{y}$ is in our power.[55] The whole matter reduces

to this, we must find the curve[56] in which the ordinate is such that

---

[54] Leibniz does not see that this result immediately gives him the equation that he requires. Thus $x = c\,\mathrm{Log}\,y$, as he would have written it; the usual omission of the arbitrary constant does not matter in this case, so long as BA is taken as unity, which is possible with Leibniz's data.

[55] Here he seems to recognize that he has the solution. The next sentence is, however, very strange. As long ago as Nov. 1675 he has written $\int a^2/y$ as Log y, and recognized the connection between the integral and the quadrature of the hyperbola; and yet he says "unless I am mistaken, $\int dy/y$ is always in our power." Now notice that in the date there is no day of the month given, contrary to the usual custom with these manuscripts so far; can it be possible that this date was afterward added from memory, and that the manuscript should bear an earlier date? If not we must conclude that Leibniz has not yet attained to a correct idea of the meaning of his integral sign, and is still worried by the necessity (as it appears to him) of taking the y's in arithmetical progression.

[56] The passage in the original Latin is very ambiguous, and it may be that it is not quite correctly given; I think, however, that I have given the correct idea of what Leibniz intended. One has to draw an *auxiliary* curve, in which $y = dy/dx$, and then find its area; in that case it should be "divided by the differences of the abscissae" instead of "divided by the abscissae."

it is equal to the differences of the ordinates divided by the abscissae, and then find the quadrature of that figure.

$$d\,\overline{\sqrt{ay}} = \frac{1}{\sqrt{ay}} \qquad (57)$$

Figures of this kind, in which the ordinates are dy/y, dy/y^2 dy/y^3, are to be sought in the same way as I have obtained those whose ordinates are $y\,dy$, $y^2\overline{dy}$, etc. Now $w/a=\overline{dy}/y$, and since $\overline{dy}$ may be taken to be constant and equal to β,[58] therefore the curve, in which $w/a=\overline{dy}/y$, will give $wy=a\beta$, which would be a hyperbola.[59]   Hence the figure, in which $dy/y=z$, is a hyperbola, no matter how you express y, and if y is expressed by ϕ^2 we have $dy=2\phi$, and $\dfrac{2\phi}{\phi^2} = \dfrac{2}{\phi}$.   Now, $c\displaystyle\int \frac{dy}{y} = \frac{a}{f}\,z$, and therefore $\dfrac{fc}{a}\displaystyle\int \frac{\overline{|1}}{y} = z$, which thus appertains to a logarithm.[60]

Thus we have solved all the problems on the inverse method of tangents,[61] which occur in Vol. 3 of the Correspondence of Descartes, of which he solved one himself, as he says on page 460, letter 79, Vol. 3; but the solution is not given; the other he tried to solve but could not, stating that it was an irregular line, which in any case was not in human power, nay not within the power of the angels unless the art of describing it is determined by some other means.

§ 13.

This manuscript bears no date: however, it was probably written very shortly after his call on Hudde at Amsterdam, on his way home from England (the second visit)

[57] An interpolated note, marking a sudden thought or guess; for the next sentence carries on the train of thought that has gone before. Query, some interval of time, either short (such as for a meal) or long (continued the next day), may have occurred here.

[58] This cannot be referred back to the present problem, since Leibniz has already assumed in it that dz and dx are constant. This may account for the fact that he has hesitated to say that the integral represents a logarithm. .

[59] This working is intended to apply to the auxiliary curve mentioned above, w standing for dx, and β for dy; hence the curve is not a hyperbola; Leibniz seems to have been misled by the appearance of the equation suggesting $xy =$ constant.

[60] Here apparently he leaves the muddle, in which he has entangled himself, and returns to his original equation; he then remembers that he has found before that the integral in question leads to a logarithm. (See p. 71.)

[61] He has not solved either of them; nor can it be said from this that "Leibniz in 1676 sought and found the curve whose subtangent is constant." Of all the work that Leibniz has done hitherto, there is none that is so inconclusive as this in comparison.

to Hanover. Leibniz stayed in Holland from October 1676 to December of that year; hence the date may be fairly accurately assigned.

Hudde showed me that in the year 1662 he already had the quadrature of the hyperbola, which I found was the very same as Mercator also had discovered independently, and published. He showed me a letter written to a certain van Duck, of Leyden I think, on this subject. His method of tangents is more complete than that of Sluse, in that he is able to use any arithmetical progression, as in a simple equation, whereas Sluse and others can use only one. Hence constructions can be made simple, while terms can be eliminated at will. This also can be made use of for eliminating any letter with greater facility, for numerous equations of all sorts are thereby rendered fit for elimination.

$$x^3 + px^2 + qx = 0 \qquad\qquad x^2 + xy + y^2 + x + y + a = 0$$
$$\begin{array}{c} y \qquad\qquad \dfrac{y}{y^2} \cdot \dfrac{y}{y^2} \\ y^3 \end{array} \qquad\qquad 2x\overline{dx} + x\overline{dy} + 2y\overline{dy} + \overline{dx} + \overline{dy} = 0$$
$$\overline{ydx}$$
$$3x^2 + 2px^2 + qx\ 0 \qquad\qquad \frac{t}{y} = \frac{\overline{dx}}{dy} = \frac{x+2y+1}{y+2x+1}$$
$$2yx^2 + yx$$
$$y^2x$$

What I had observed with regard to triangular numbers for three equal roots, and pyramidal numbers for four, was already known to him, and indeed even more generally,

$$\begin{array}{rrrrrrrrr} -1 & 0 & 1 & 2 & 3 & 4 & 5 & 6 \\ -3 & -1 & 0 & 0 & 1 & 3 & 6 & 10 & 15 \\ -4 & -1 & 0 & 0 & 0 & 1 & 4 & 10 & 20 \end{array}$$

Here it must be observed that the number of zeros increases, as this is of the greatest service in separating roots.

He has also rules for multiplying equations, so that they are not only determined for equal roots, but also for roots increasing arithmetically, or geometrically, or according to any progression.

Hudde has a most elegant construction for describing two curves, one outside and the other inside a circle, which are capable of quadrature, and by means of these curves he finds the true area of a circle so nearly, that with the help of the dodecagon, in a number of six figures, there is an error of only three units, or 3/100000.

He has a method for finding the real roots of equations, having some roots real and the rest impossible, by the help of another equation having all its roots real, and as many in number as he previously had of real and impossible together.

He had an example of a beautiful method of finding sums of series by the continuous subtractions of geometrical progressions. He subtracts geometrical progressions whose sums are also geometrical progressions, and thus he can find the sums of the sums, and so he obtains the sum of the series. This method is excellent for a series whose numerators are arithmetical, and denominators geometrical, such as

$$\frac{1}{2}, \; \frac{2}{4}, \; \frac{3}{8}, \; \frac{4}{16}, \; \ldots\ldots\ldots \; .$$

He has three series, like those of Wallis, for interpolations for the circle. He says that there are no more by that method, I think.

Also he can very often write down the quadratures of irrationals, as also their tangents, without eliminating irrationals, or fractions, etc.

§ 14.

November, 1676.

Calculus Tangentium differentialis.
[Differential calculus of tangents.]

$$\overline{dx} = 1, \quad \overline{dx^2} = 2x, \quad \overline{dx^3} = 3x^2, \quad \text{etc.}$$

$$\overline{d\frac{1}{x}} = -\frac{1}{x^2}, \quad \overline{d\frac{1}{x^2}} = -\frac{2}{x^2}, \quad \overline{d\frac{1}{x^3}} = \frac{3}{x^2}, \quad \text{etc.}$$

$$\overline{d\sqrt{x}} = \frac{1,}{\sqrt{x}} \quad \text{etc.}$$

From these the following general rules may be derived for the differences and sums of the simple powers:

$$\overline{dx^e} = e, x^{e-1}, \text{ and conversely } \int \overline{x^e} = \frac{x^{e+1}}{e+1}.$$

Hence, $\overline{d\frac{1}{x^2}} = \overline{dx^{-2}}$ will be $-2x^{-3}$ or $-\frac{2}{x^3}$,

and $\overline{d\sqrt{x}}$ or $dx^{\frac{1}{2}}$ will be $-\tfrac{1}{2}x^{-\frac{1}{2}}$ or $-\tfrac{1}{2}\sqrt{\frac{1}{x}}$.

Let $y = x^2$, then $\overline{dy} = 2x\,\overline{dx}$ or $\overline{\frac{dy}{dx}} = 2\,x$.

This reasoning is general, and it does not depend on what the progression for the x's may be.[62] By the same method, the general rule is established as:

$$\frac{\overline{dx^e}}{dx} = e\,x^{e-1}, \text{ and } \int \overline{x^e\,\overline{dx}} = \frac{x^{e+1}}{e+1}.$$

Suppose that we have any equation whatever, say,

$$ay^2 + byx + cz^2 + f^2x + g^2y + h^3 = 0,$$

and suppose that we write $y+dy$ for y, and $x+dx$ for x, we have, by omitting those things which should be omitted, another equation

$$\left.\begin{array}{c} ay^2 + byx + cx^2 + f^2x + g^2y + h^3 = 0 \\ \hline a2\overline{dyy} + by\overline{dx} + 2cx\overline{dx} + f^2\overline{dx} + g^2\overline{dy} \\ bx\overline{dy} \\ \hline a\,\overline{dy^2} + b\,\overline{dxdy} + c\,\overline{dx^2} = 0 \end{array}\right\} = 0 \qquad (63)$$

This is the origin of the rule published by Sluse. It can be extended indefinitely: Let there be any number of letters, and any formula composed from them; for example, let there be the formula made up of three letters,

$$ay^2 \quad bx^2 \quad cz^2 \quad fyx \quad gyx \quad hxz \quad ly \quad mx \quad nz \quad p=0.$$

From this we get another equation

ay^2	bx^2	cz^2	fyx	simi-	ly	mx	simi-	p
$2a\overline{dyy}$	$2b\overline{dxx}$	$2c\overline{dzz}$	$fy\overline{dx}$	larly	$l\overline{dy}$	$m\overline{dx}$	larly	
			$fx\overline{dy}$					
$a\,\overline{dy^2}$	$b\overline{dx^2}$	$c\overline{dz^2}$	$f\overline{dxdy}$	.....				

It is plain from this that by the same method tangent planes

[62] AT LAST! The recognition of the fact that neither dx nor dy need necessarily be constant, and the use of another letter to stand for the function that is being differentiated, mark the beginning, the true beginning, of Leibniz's development of differentiation. Later in this manuscript we find him using the third great idea, probably suggested by the second of those given above, namely, the idea of substitution, by means of which he finally attains to the differentiation of a quotient, and a root of a function.

It is very suggestive that this remarkable advance occurs after his second visit to London, while he is staying in Holland. Did some one tell him then of the work of Newton, or of Barrow's method (which is geometrically an exact equivalent of substitution), pointing out those things of which he had **not** perceived the drift, or is it the result of his intercourse with Hudde? For the date is that of his stay at The Hague. See Chap. VI, "Leibniz in London."

[63] This is Barrow all over; even to the words *omissis omittendis* instead of Barrow's *rejectis rejiciendis*. Lect. X, Ex. 1 on the differential triangle at the end of the lecture.

to surfaces may be obtained, and in every case that it does not matter whether or no the letters x, y, z have any known relation, for this can be substituted afterward.

Further, the same method will serve admirably, even though compound fractions or irrationals enter into the calculation, nor is there any need that other equations of a higher degree should be obtained for the purpose of getting rid of them; for their differences are far better found separately and then substituted; hence the ordinary method of tangents will not only proceed when the ordinates are parallel, but it can also be applied to tangents and anything else, ay, even to those things that are related to them, such as proportions of ordinates to curves, or where the angle of the ordinates changes according to some determined law. It will be worth while especially to apply the method to irrationals and compound fractions.[64]

$$d\overline{\sqrt[2]{a+bz+cz^2}}. \quad \text{Let } a+bz+cz^2 = x;$$

then
$$d\overline{\sqrt[2]{x}} = -\frac{1}{2\sqrt{x}}, \text{ and } \frac{dx}{dz} = b + 2cz;$$

therefore
$$d\overline{\sqrt[2]{a+bz+cz^2}} = -\frac{b+2cz}{2\,dz\,\overline{\sqrt{a+bz+cz^2}}}$$

Taking any equation between two letters x and y for a curve, and determining the equation of the tangent, either of the two letters x or y can be eliminated, so that all that remains is the other together with $\overline{dx}$ and $\overline{dy}$; and this will be worth while doing in all cases to facilitate the calculation.

If three letters are given, say x, y and z, and the value of $\overline{dz}$ is expressed in terms of x or y (or even of both), an equation for the tangents will at length be obtained, in which again there will be left only one or other of the letters x or y together with the two, $\overline{dx}$ and $\overline{dy}$; sometimes z itself cannot be eliminated. Also this can be deduced in all cases of an assumed value of $\overline{dz}$, and in the same way more additional letters can be taken. Thus, bringing together every general calculus into one, we obtain the most general of them all. Besides, the assumption of a large number of letters may be employed to solve problems on the inverse method of tangents, with the assistance of quadratures.

[64] Here we have the idea of substitutions, which made the Leibnizian calculus so superior to anything that had gone before. Note that he still has the erroneous sign that he obtained for the differentiation of $\sqrt{x}$ at the beginning of this manuscript. Also that the dz is wrongly placed in the denominator of the result.

Thus, if the following problem is set for solution: It is given that the sum of the straight lines CB, BP or

$$y + y\frac{\overline{dy}}{dx} = xy \; ;$$

we have

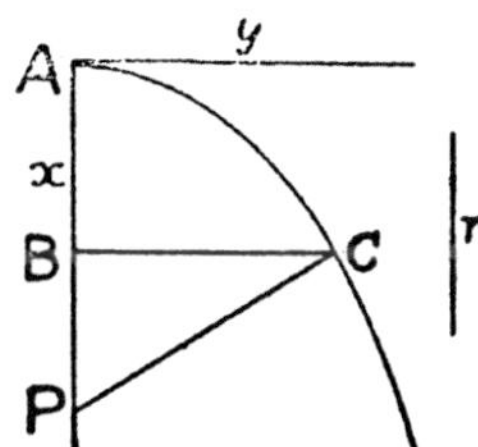

$$\overline{dx} + \overline{dy} = x\,dx$$

$$\text{or } x + y = \frac{x^2}{2} \; .$$

Thus we have the curve in which the sum of CB + BP (multiplied by a constant r) is equal to the rectangle AB.BC.

[There are two marginal notes by Leibniz that must be referred to, in this manuscript. The first reads:

"It is especially to be observed about my calculus of differences that, if

$$b, y\,dx + x\,dy + \text{etc.} = 0$$

then $byx + \int$ etc. $= 0$, and so on for the rest. It is to be seen what is to be done about the h^3. For the purpose of making these calculations better, the equation $ay^2 + byx + cx^2 +$ etc. can be changed into something else by means of another relation of the curve, and if it turns out all right it may be compared with another calculation of the differences, since it comes to the thing as by the first."

The two points to be noticed are that Leibniz now for the first time recognizes the need of considering the arbitrary constant of integration, though he hardly grasps how it arises, and that even now he cannot refrain from harking back to his obsession of the obtaining of several equations for comparison. This note is not made any the easier to understand by its being starred by Gerhardt for reference to the differentiation of x^2, whereas it obviously (when you come later to the passage) refers to the differentiation of the equation of the second degree.

The second note refers to the substitution of $x + dx$ for x and $y + dy$ for y, and reads:

"Either dx or dy can be expressed arbitrarily, a new equation being obtained; and either dx or dy being taken away, x, or y, say, can be otherwise expressed in terms of the quantities. It is not true, I think, that this is so, for then a catalogue of all curves capable of quadrature would result, by supposing one or other of them to be constant."

The point to be noticed in this rather ambiguous statement is that Leibniz is still thinking of his catalogue, and is not himself convinced of the completeness of his method for all purposes.]

§ 15.

There is an interval of nearly seven months between the date of the manuscript last considered and the one that now follows. This interval has been full of work; for we now find a clear exposition of the rules for the differentiation of a sum, difference, product, quotient, etc., though these are without proof, or indication of the manner in which they have been obtained. There is also no rule given for a logarithm, an exponential, or a trigonometrical ratio. Leibniz may have known them, but even then it would not be surprising to find them left out; for Leibniz's great idea was the use of his method to facilitate calculation. We must conclude therefore that these rules are a development of the method of substitution outlined in the preceding manuscript.

This essay has several peculiar characteristics of its own, which distinguish it from those that have gone before. It is written throughout in French; it is to some extent historical and critical, having the appearance of being prepared for publication, or possibly as a letter; this is corroborated by the fact that there is an original draft and a more fully detailed revision. Could it be that this is the original of Leibniz's communication of this method to Newton and others? If so, Leibniz is very careful not to give much away. The figures are strongly reminiscent of Barrow, but the context does not deal with subtangents, which are such a feature in all Barrow's work.

The start from the work of Sluse is peculiar; it seems to suggest that Leibniz is pointing out that his method is a fuller development of that of the former. Leibniz has already hazarded two different guesses at the origin of the rules given by Sluse; the second, namely, by substitution of $x + dx$ for x, etc., being the more probable. Is Leibniz trying to draw a red herring across the trail, the real trail that leads to Barrow's a and e?

11 July 1677.

Méthode générale pour mener les touchantes des Lignes Courbes sans calcul, et sans réduction des quantités irrationelles et rompues.

[General method for drawing tangents to curves without calculation,and without reducing irrational or fractional quantities.]

Slusius has published his method of finding tangents to curves
without calculation, in which the equation is purged of irrational
or fractional quantities.

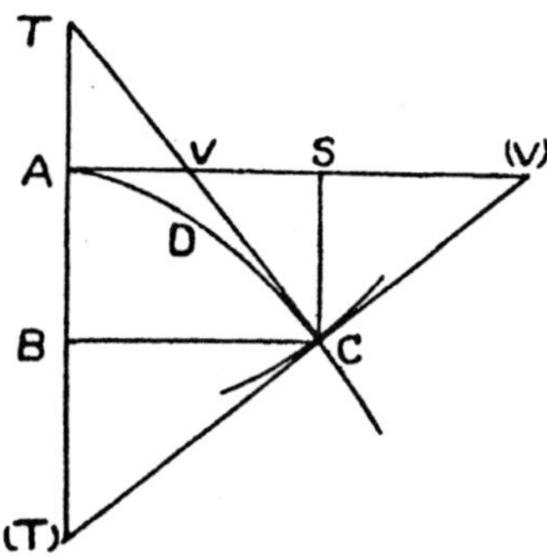

For example, a curve DC being given, in which the equation
expresses the relation between BC and AS, which we will call y,
and AB or SC, which we will call x; let this be

$$a + bx + cy + dxy + ex^2 + fy^2 + gx^2y + hxy^2 + kx^3 + ly^3 + \text{etc.} = 0.$$

One has only to write

$$0 + b\xi + cv + dxv + 2ex\xi + 2fyv + gx^2v + hy^2\xi + 3kx^2\xi + 3ly^2v$$
$$dy\xi \qquad\qquad 2gxy\xi \quad 2hxyv$$

$$+ mx^2y^2 + nx^3y + pxy^3 + qx^4 + ry^4 \qquad\qquad (65)$$
$$+ 2mx^2yv + nx^3v + py^3\xi + 4qx^3\xi + 4ry^3v$$
$$+ 2mxy^2\xi + 3nx^2y\xi + 3\,py^2xv$$

that is to say, if the equation is changed to a proportion,

$$\frac{\xi}{v} = \frac{c + dx + 2fy + gx^2 + 2hxy + 3ly^2 + 2mx^2y + \text{etc.}}{b + dy + 2ex + 2gxy + hy^2 + 3kx^2 + \text{etc.}};$$

and, supposing that $\dfrac{\xi}{v}$ expresses the ratio $\dfrac{\text{TB}}{\text{BC}=x}$ or $\dfrac{\text{CS}=y}{\text{SV}}$,
then TB or SV can be obtained, if BC and SC are supposed to
be given. When the given magnitudes, b, c, d, e, etc., with their
proper signs, make the value of ξ/v a negative magnitude, the tangent will not be CT which goes toward A, the start of the abscissa
AB, but C(T) which goes away from it. That is all that has been

<hr>

[65] This line represents the "etc." of the original equation, and is set down
for the purpose of getting the derived terms; the complete derived equation
therefore consists of the two lines above and the two below. Note the omission of the negative sign, when changing from the equation to the proportion.

published up to the present time, easy to understand by any one that is versed in these matters. But when there are irrational or fractional magnitudes, which contain either x or y or both, this method cannot be used, except after a reduction of the given equation to another that is freed from these magnitudes. But at times this increases to a terrible degree the calculation and obliges us to rise to very high dimensions, and leads us to equations for which the process of depression is often very difficult. I have no doubt that the gentlemen[66] I have just named know the remedy that it is necessary to apply, but as it is not as yet in common use, and is I believe known to but a few, also because it gives the finishing touch to the problem that Descartes said was the most difficult to solve of all geometrical problems, because of its general utility, I have thought it a good thing to publish it.

Suppose we have any formula or magnitude or equation such as was given above,

$$a + bx + cy + dxy + ex^2 + fy^2 + \text{etc.} \, ;$$

for brevity let us call it ω; that which arises from it when it is treated in the manner given above, namely,

$$b\xi + cv + dxv + dy\xi + \text{ etc.} \, ;$$

will be called $d\omega$; and in the same way, if the formula is λ or μ, then the result above will be $d\lambda$ or $d\mu$, and similarly for everything else. Now let the formula or equation or magnitude ω be equal to λ/μ, then I say that $d\omega$ will be equal to $\dfrac{\mu\,\overline{d\lambda} - \lambda\,\overline{d\mu}}{\mu^2}$. This will be sufficient to deal with fractions.

Again, let ω be equal to $\sqrt[z]{\omega}$, then $d\omega = z.\sqrt[z]{\dfrac{\overline{dw}}{\omega^{\frac{z-1}{z}}}}$; and this will be sufficient for the proper treatment of irrationals.

Algorithm of the new analysis for maxima and minima, and for tangents.

Let $AB = x$, and $BC = y$, and let TVC be the tangent to the curve AC; then the ratio $\dfrac{\text{TB}}{\text{BC} = y}$ or $\dfrac{\text{SC} = x}{\text{SV}}$ will be called $\dfrac{dx}{dy}$.

---

[66] Leibniz, at the beginning, first wrote, "Hudde, Sluse, and others"; but later he struck out all but Sluse. (Gerhardt.)

Let there be two or more other curves, AF, AH, and suppose
that $BF = v$ and $BH = w$, and that the straight line FL is the tangent

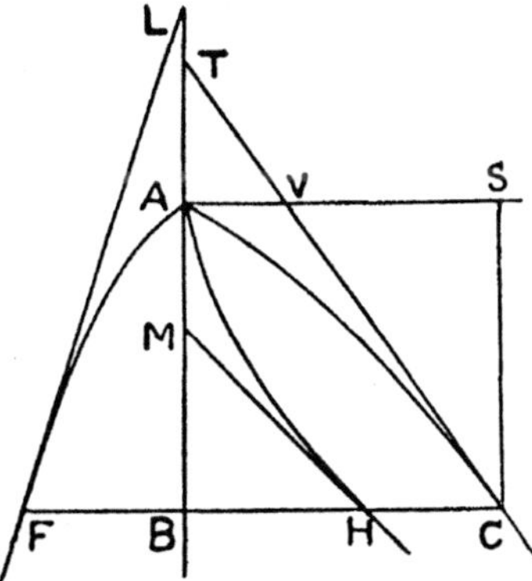

to the curve AF, and MH to the curve AH; also $\dfrac{LF}{FB} = \dfrac{dx}{dv}$, and
$\dfrac{MH}{BH} = \dfrac{dx}{dw}$; then I say that dy, or dvw, will be equal to $vdw + wdv$;
and if $v = w = x$, and $y = vw = x^2$, then by substituting x for v and
for w, we shall have $dvw = 2xdx$.

(This will also hold good if the angle ABC is either acute or
obtuse; also if it is infinitely obtuse, that is to say, if TAC is a
straight line.)

[Of this rough draft there is the following revision, and this
obviously comes within the same period. (Gerhardt.)]

Fermat was the first to find a method which could be made
general for finding the straight lines that touch analytical curves.
Descartes accomplished it in another way, but the calculation that
he prescribes is a little prolix. Hudde has found a remarkable
abridgment by multiplying the terms of the progression by those
of the arithmetical progression. He has only published it for equations in one unknown; although he has obtained it for those in two
unknowns. Then the thanks of the public are due to Sluse; and
after that, several have thought that this method was completely
worked out. But all these methods that have been published suppose that the equation *has been reduced* and cleared of fractions
and irrationals; I mean of those in which the variables occur. I
however have found means of obviating these useless reductions,
which make the calculation increase to a terrible degree, and oblige
us to rise to very high dimensions, in which case we have to look

for a corresponding depression with much trouble; instead of all this, everything is accomplished at the first attack.

This method has more advantage over all the others that have been published, than that of Sluse has over the rest, because it is one thing to give a simple abridgment of the calculation, and quite another thing to get rid of reductions and depressions. With respect to the publication of it, on account of the great extension of the matter which Descartes himself has stated to be the most useful part of Geometry, and of which he has expressed the hope that there is more to follow—in order to explain myself shortly and clearly, I must introduce some *fresh characters,* and give to them a *new Algorithm,* that is to say, altogether special rules, for their addition, subtraction, multiplication, division, powers, roots, and also for equations.

Explanation of the characters.

Suppose that there are several curves, as CD, FE, HJ, connected with one and the same axis AB by ordinates drawn through one and the same point B, to wit, BC, BF, BH. The tangents CT, FL, HM to these curves cut the axis in the points T, L, M; the

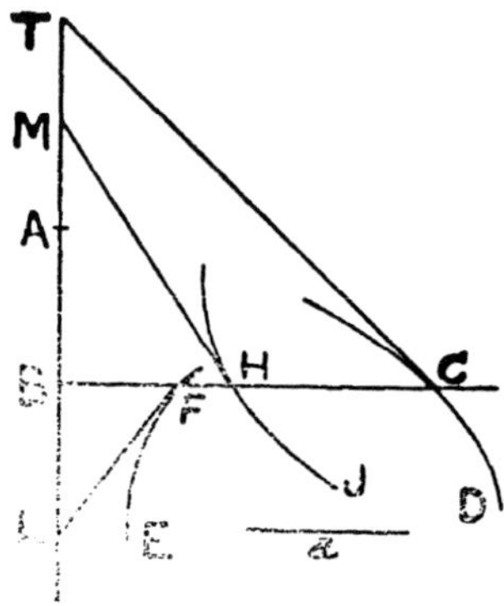

point A in the axis is fixed, and the point B changes with the ordinates. Let $AB = x$, $BC = y$, $BF = w$, $BH = v$; also let the ratio of TB to BC be called that of dx to dy, and the ratio of LB to BF that of dx to dw, and the ratio of MB to BH that of dx to dv. Then if, for example, y is equal to vw, we should say dvw instead of dy, and so on for all other cases. Let a be a constant straight line; then, if y is equal to a, that is, if CD is a straight line parallel to AB, dy or da will be equal to 0, or equal to zero. If the magnitude dx/dw comes out negative, then FL, instead of being drawn toward A, above B, will be drawn in the contrary direction, below B.

Addition and Subtraction. Let $y = v \pm w(\pm)a$, then $\overline{dy}$ will be equal to $\overline{dv} \pm \overline{dw}(\pm)0$.

Multiplication. Let y be equal to avw, then $\overline{dy}$ or $\overline{davw}$ or $a\,\overline{dvw}$ will be equal to $av\,\overline{dw} + aw\,\overline{dv}$.

Division. Let y be equal to $\dfrac{v}{aw}$, then $\overline{dy}$ or $d\dfrac{v}{aw}$

or $\dfrac{1}{a}\,d\dfrac{v}{w}$ will be equal to $\dfrac{w\,\overline{dv} - v\,\overline{dw}}{aw^2}$.

The rules for *Powers* and *Roots* are really the same thing.

Powers. If $y = w^z$, (where z is supposed to be a certain number), then $\overline{dy}$ will be equal to $z,\ w^{z-1},\ dw$.

Roots or extractions. If $y = \sqrt[z]{w}$, then $\overline{dz} = z\,\dfrac{dw}{\sqrt[\frac{z}{w}]{}}$.

Equations expressed in rational integral terms.

$$a + bv + cy + tvy + ev^2 + fy^2 + gv^2y + hvy^2 + kv^3 + ly^3$$
$$+ mv^2y^2 + nv^3y + pvy^3 + qv^4 + ry^4 = 0,$$

supposing that a, b, c, t, e, etc. are magnitudes that are known and determined; then we should have

$$0 = b\overline{dv} + c\overline{dy} + tv\overline{dy} + 2ev\overline{dv} + 2fy\overline{dy} + gv^2\overline{dy} + hy^2\overline{dv}$$
$$ty\overline{dv} \qquad\qquad +2gvy\overline{dy} +2hvy\overline{dy}$$

$$+ 3ly^2\overline{dy} + 2mv^2y\overline{dy} + nv^3\overline{dy} + py^3\overline{dv} + 4qv^3\overline{dv} + 4ry^3\overline{dy}$$
$$+ 2mvy^2\overline{dv} + 3nv^2y\overline{dv} + 3py^2v\overline{dy}$$

This rule can be proved and continued without limit by the preceding rules; for, if

$$a + bv + cy + tvy + ev^2 + fy^2 + gv^2y + \text{etc.} = 0,$$

then $da + dbv + dcy + tdvy + edv^2 + fdy^2 + gdv^2y + $ etc. will also be equal to 0. Now $da = 0$, $dbv = bdv$, $dcy = cdy$, $dvy = vdy + ydv$; also $dv^2 = 2vdv$, since dv^z is equal to z, v^{z-1}, dv, that is to say (by substituting 2 for z) $2vdv$; and $dv^2y = v^2dy + 2vydv$, for, supposing that $w = v^2$, then dv^2y will be dwy, and $dwy = ydw + wdy$, and dw or $dv^2 = 2vdv$; hence in the value of dwy, substituting for w and dw the values found for them, we shall have $dv^2y = v^2dy + 2vydv$, as obtained above. This can go on without limit. If in the given equation $a + bv + cy$ + etc. $= 0$, the magnitude v were equal to x, that is to say if the line JH were a straight line which when produced passed through the point A, making an angle of 45 degrees with the axis, then the resulting equation, transformed into a proportion, would give the rule for the method of tangents, as published by Sluse; and, in consequence, this is nothing but a particular case or corollary of the general method.

Equations complicated in any manner with fractions and irrationals. These could be treated in the same way without any calculation, by supposing that the denominator of the fraction or the magnitude of which it is necessary to take the root is equal to a magnitude or letter, which is to be treated according to the preceding rules.[67]

Also, when there are magnitudes which have to be multiplied by one another, there is no need to make this multiplication in reality, which saves still more labor. One example will be sufficient.

[No example is given, however; but the following seems to have been added later, according to Gerhardt.]

Lastly this method holds good when the curves are not purely analytical, and even when their nature is not expressed by such ordinates, and in addition it gives a marvelous facility for making geometrical constructions. The true reason for an abridgment so admirable, and one that enables us to avoid reductions of fractions and irrationals, is that one can always make certain, by means of the preceding rules, that the letters dy, dv, dw, and the like, shall not occur in the denominator of the fraction, or under the root-sign.

§ 16.

The next manuscript appears to be a more detailed revision of the one last considered. It bears no date; but it is safe to say that it belongs to a considerably later period than that of July 1677. For in this are given, by means of the *infinitely small* quantities dx and dy, proofs of the

---

[67] The complete statement of the method of substitutions.

fundamental rules for the first time; the figure notation is changed from the clumsy C, (C), ((C)) to the neat $_1C$, $_2C$, $_3C$; the notation for proportion is now $a:b::c:d$; and there are several other changes that readers will notice as they go along. The ideas of Leibniz are now approaching crystallization, as is evidenced by the fact that $\int y\,dx$ is clearly stated for the first time to be the sum of *rectangles made from y and dx*. It is rather astonishing, however, in this connection to find $\int \overline{x+y-v} = \int x + \int y - \int v,$ which can have no significance according to the above definition; and also to find the whole thing explained by arithmetical series, in which however it is to be observed that dx is not taken to be constant. But for this one might almost place this later than the publication of the method in the *Acta Eruditorum* in 1684; in this essay Leibniz gave a full account of his rules without proofs, and is evidently trying to get away from the idea of the infinitely small, an effort which culminates in the next, and last, manuscript of this set.

If then we guess the date to be about 1680, probably we shall not be very far out.

A remarkable feature of this manuscript is the omission of really necessary figures, without which the text is very hard to follow. Of course this manuscript was written for publication, and the suggestion may be made that the diagrams were drawn separately, just as in books of that time they were printed separately on folding plates; but then, why has he given three diagrams? The only other suggestion that can be made as far as I can see is that he was referring to texts, in which the diagrams were already drawn, by Gregory St. Vincent, Cavalieri, James Gregory (one of whose theorems he quotes), Barrow (who strangely enough also quotes the very same theorem), Wallis, and others. For he mentions many of these authors, but there is never a word about Barrow. I consider that he was looking up their theorems to show *how much superior his method was to any of theirs.*

It is to be observed that not even in this manuscript is there any mention of logarithms, exponentials, or trigonometrical ratios. We shall see later that Leibniz is reduced to obtaining the integral of $(a^2 + x^2)^{1/2}$ by reference to a figure and its quadrature; that is to say, he is apparently unable to perform the integration analytically. It therefore follows that, if he got a great deal from Barrow, he was unable to understand the Lect. XII, App. I of the *Lectiones Geometricae.*

The final conclusion that I personally have come to, after completing this examination of the manuscripts of Leibniz, as far as they are given by Gerhardt is this:

As far as the actual invention of the calculus as he understood the term is concerned, Leibniz received no help from Newton or Barrow; but for the ideas which underlay it, he obtained from Barrow a very great deal more than he acknowledged, and a very great deal less than he would like to have got, or in fact would have got if only he had been more fond of the geometry that he disliked. For, although the Leibnizian calculus was at the time of this essay far superior to that of Barrow on the question of useful application, it was far inferior in the matter of completeness.

(No date.)

Elementa calculi novi pro differentiis et summis, tangentibus et quadraturis, maximis et minimis, dimensionibus linearum, superficierum, solidorum, aliisque communem calculum transcendentibus.

[The elements of the new calculus for differences and sums, tangents and quadratures, maxima and minima, dimensions of lines, surfaces, and solids, and for other things that transcend other means of calculation.]

Let CC be a line, of which the axis is AB, and let BC be ordinates perpendicular to this axis, these being called y, and let AB be the abscissae cut off along the axis, these being called x.

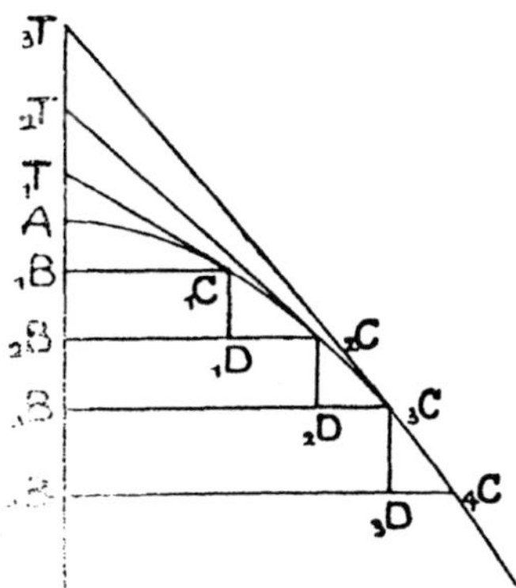

Then CD, the differences of the abscissae, will be called dx; such are $_1C_1D$, $_2C_2D$, $_3C_3D$, etc. Also the straight lines $_1D_2C$, $_2D_3C$, $_3D_4C$, the differences of the ordinates, will be called dy. If now these dx and dy are taken to be infinitely small, or the two points on the curve are understood to be at a distance apart that is less than any given length, i. e., if $_1D_2C$, $_2D_3C$, etc. are considered as the momentaneous increments[68] of the line BC, increasing continuously as it descends along AB, then it is plain that the straight line joining these two points, $_2C_1C$ say, (which is an element of the curve or a side of the infinite-angled polygon that stands for the curve), when produced to meet the axis in $_1T$, will be the tangent to the curve, and $_1T_1B$ (the interval between the ordinate and the tangent, taken along the axis) will be to the ordinate $_1B_1C$ as $_1C_1D$ is to $_1D_2C$; or, if $_1T_1B$ or $_2T_2B$, etc. are in general called t, then $t:y :: dx:dy$. Thus to find the differences of series is to find tangents.

For example, it is required to find the tangent to the hyperbola. Here, since $y = \dfrac{aa}{x}$, supposing that in the diagram, x stands for AB the abscissa along an asymptote, and a for the side of the power, or of the area of the rectangle AB.BC; then

$$dy = -\frac{aa}{xx}dx,$$

[68] Leibniz has evidently seen Newton's work at the time of this composition; also the use of the word "descends" in the next line again suggests Barrow, while the figure is exactly like the top half of the diagram given by Barrow for Lect. XI, 10, which is the theorem of Gregory that is quoted by Leibniz also. For this figure, see Note 71, p. 140.

as will be soon seen when we set forth the method of this calculus;

hence $dx : dy$ or $t : y :: -xx : aa :: -x : \dfrac{aa}{x} :: -x : y$; therefore $t = -y$,

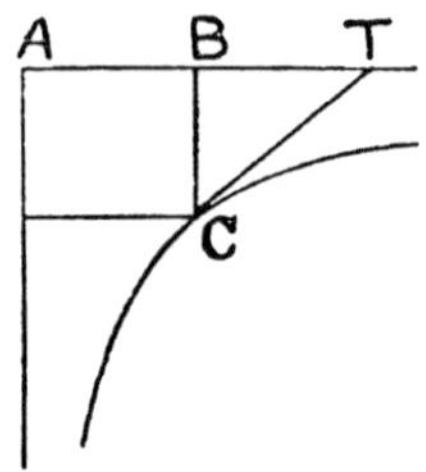

that is, in the hyperbola BT will be equal to AB, but on account of the sign $-x$, BT must be taken not toward A but in the opposite direction.

Moreover, differences are the opposite to sums; thus $_4B\,_4C$ is the sum of all the differences such as $_3D\,_4C$, $_2D\,_3C$, etc. as far as A, even if they are infinite in number. This fact I represent thus, $\int dy = y$. Also I represent the area of a figure by the sum of all the rectangles contained by the ordinates and the differences of the abscissae, i. e., by the sum $_1B\,_1D + _2B\,_2D + _3B\,_3D +$ etc. For the narrow triangles $_1C\,_1D\,_2C$, $_2C\,_2D\,_3C$, etc., since they are infinitely small compared with the said rectangles, may be omitted without risk; and thus I represent in my calculus the area of the figure by $\int y\,dx$, or the sum of the rectangles contained by each y and the dx that corresponds to it; here, if the dx's are taken equal to one another, the method of Cavalieri is obtained.

But we, now mounting to greater heights, obtain the area of a figure by finding the figure of its summatrix or quadratrix; and of this indeed the ordinates are to the ordinates of the given figure in the ratio of sums to differences; for instance, let the curve of the figure required to be squared be EE, and let the ordinates to it, EB, which we will call e, be proportional to the differences of the ordinates BC, or to dy; that is let $_1B\,_1E : _2B\,_2E :: _1D\,_2C : _2D\,_3C$, and so on; or again, let $A\,_1B : _1B\,_1C$, $_1C\,_1D : _1D\,_2C$, etc., or $dx : dy$ be in the ratio of a constant or never-varying straight line a to $_1B\,_1E$ or e; then we have

$$dx : dy :: a : e, \text{ or } e\,dx = a\,dy;$$
$$\therefore \int e\,dx = \int a\,dy.$$

But $e\,dx$ is the same as e multiplied by its corresponding dx, such as the rectangle $_3B\,_4E$, which is formed from $_3B\,_3E$ and $_3B\,_4B$; hence, $\int e\,dx$ is the sum of all such rectangles, $_3B\,_4E + _2B\,_1E + _3B\,_2E$ + etc., and this sum is the figure $A\,_4B\,_4EA$, if it is supposed that the

dx's, or the intervals between the ordinates e, or BC, are infinitely small. Again, $a\,dy$ is the rectangle contained by a and dy, such as is contained by $_3D\,_4C$ and the constant length a, and the sum of

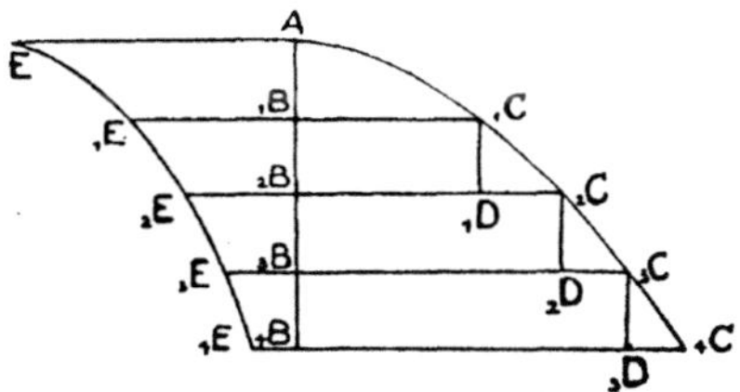

these rectangles, namely $\int a\,dy$, or $_3D\,_4C.a+_2D\,_3C.a+_1D\,_2C.a+$ etc. is the same as $_3D\,_4C+_2D\,_3C+_1D\,_2C+$ etc. into a, that is, the same as $_4B\,_4C.a$; therefore we have $\int a\,dy=a\int dy=ay$. Therefore $\int e\,dx=ay$, that is, the area $A\,_4B\,_4EA$ will be equal to the rectangle contained by $_4B\,_4C$ and the constant line a, and generally ABEA is equal to the rectangle contained by BC and a.[69]

Thus, for quadratures it is only necessary, being given the line EE, to find the summatrix line CC, and this indeed can always be found by calculus, whether such a line is treated in ordinary geometry or whether it is transcendent and cannot be expressed by algebraical calculation; of this matter in another place.

Now the triangle for the line I call the characteristic of the line, because by its most powerful aid there can be found theorems about the line which are seen to be admirable, such as its length, the surface and solid produced by its rotation, and its center of gravity; for $_1C\,_2C$ is equal to $\sqrt{dx.dx+dy.dy}$. From this we have

[69] Leibniz does not give a diagram, but it is not difficult to construct his figure from the enunciation that he gives for it. The whole of this paragraph should be compared with the following extract from Barrow (Lect. XI, 19), piece by piece.

"Again, let AMB be a curve of which the axis is AD and let BD be perpendicular to AD; also let KZL be another line such that, when any point M is taken in the curve AB, and through it are drawn MT a tangent to the curve AB, and MFZ parallel to DB, cutting KZ in Z and AD in F, and R is a line of given length, TF : FM = R : FZ. Then the space ADLK is equal to the rectangle contained by R and DB.

For, if DH = R and the rectangle BDHI is completed, and MN is taken to be an indefinitely small arc of the curve AB, and MEX, NOS are drawn parallel to AD; then we have
NO : MO = TF : FM = R : FZ;
NO.FZ = MO.R and FG.FZ = ES.EX.
Hence, since the sum of such rectangles as FG.FZ differs only in the least degree from the space ADLK, and the rectangles ES.EX form the rectangle DHIB, the theorem is quite obvious."

at once a method for finding the length of a curve by means of some quadrature; e. g., in the case of the parabola, if $y=\dfrac{xx}{2a}$, then we have $dy=\dfrac{xdx}{a}$, and hence $_1C\ _2C=\dfrac{dx}{a}\sqrt{aa+xx}$; hence, $_1C\ _2C:dx$ as the ordinate of the hyperbola $\sqrt{aa+xx}$ is to the constant line a; that is, $\dfrac{1}{a}\int dx\sqrt{aa+xx}$, a straight line equal to the arc of a parabola, depends on the quadrature of the hyperbola, as has already been found by others; and thus we can derive by the calculus all the most beautiful results discovered by Huygens, Wallis, van Huraet, and Neil.[70]

I said above that $t:y::dx:dy$; hence we have $t\,dy=y\,dx$, and therefore $\int t\,dy=\int y\,dx$. This equation, enunciated geometrically, gives an elegant theorem due to Gregory,[71] namely that, if BAF is a right angle, and AF = BG, and FG is parallel to AB and equal to BT, that is, $_1F\,_1G=\,_1B\,_1T$, then $\int t\,dy$, or the sum of the rectangles contained by t (e. g., $_4F\,_4G$ or $_4B\,_4T$) and dy ($_3F\,_4F$ or $_3D\,_4C$) is equal to the rectangles $_4F\,_3G+\,_3F\,_2G+\,_2F\,_1G+$ etc.; or the area of the

[70] All the things given are to be found in Barrow, but his name is not even mentioned.

[71] This is the strangest coincidence of all! For, Barrow also quotes this very same theorem of Gregory, and no other theorem; also it occurs in this very same Lect. XI that has been referred to already! *Leibniz does not give a diagram; nor from his enunciation could I complete the figure required, until I had referred to the figure given by Barrow!!!* The two diagrams are given below for comparison, Barrow's figure being the one referred to in the note above. Query, is Leibniz's figure taken from Gregory's original, which I have not been able to see, or is it the Leibnizian variation of Barrow's?

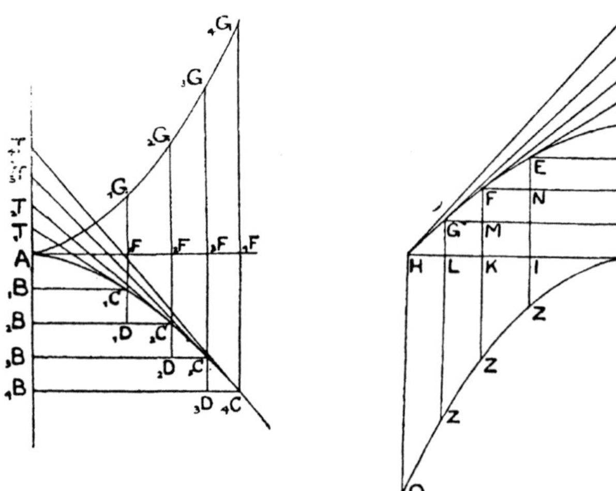

figure A$_4$F$_4$GA is equal to $\int y\,dx$, that is, to the figure A$_4$B$_4$CA; or generally, the figure AFGA is equal to the figure ABCA.

Again, other things, which are immediately evident on inspection, from a figure, are readily deduced by the calculus; for instance, in the case of the trilinear figure ABCA, the figure ABCA together with its complementary figure AFCA is equal to the rectangle ABCF, for the calculus readily shows that $\int y\,dx + \int x\,dy = xy$.

If it is required to find the volume of the solid formed by rotation round an axis, it is only necessary to find $\int y^2\,dx$; for the solid formed by a rotation round the base, $\int x^2\,dy$; for the moment about the vertex, $\int yx\,dx$; and these things serve to find the center of gravity of a figure, and also give the frusta of Gregory St. Vincent, and all that Pascal, Wallis, De Laloubère, and others have found out about these matters.

For, if it is required to find the centers of lines, or the surfaces generated by their rotation, e. g., the surface generated by the rotation of the line AC about AB, it is only necessary to find

$$\int y \sqrt{dx.\,dx + dy.\,dy}$$

or the sum of every PC applied to the axis at the point B that corresponds to it, (thus $_2$P$_2$C will be applied perpendicular to the axis AB at $_2$B), producing in this way a figure of which the above represents the area. Thus the whole thing will immediately reduce to the quadrature of some plane figure, if, instead of y and dy, their values, obtained from the nature of the ordinates and the tangents to the curve, are substituted. Thus, in the case of the parabola, if y is equal to $\sqrt{2ax}$, then $dy = \dfrac{a\,dx}{y}$ (as will be seen directly); hence we get

$$\int y\sqrt{dx\,dx + \frac{aa}{yy}\,dx\,dx} \quad\text{or}\quad \int dx\sqrt{yy + aa} \quad\text{or}\quad \int dx\sqrt{2ax + aa},$$

which depends on the quadrature of the parabola (for every $\sqrt{2ax + aa}$ or PC can be applied to a parabola, if it is supposed that AC is the parabola, and AB its axis, provided in that case the figure is changed and the curve turns its concavity toward the axis);[72] and this may be obtained by ordinary geometry, and there-

---

[72] The Latin here is rather ambiguous; query, a misprint. But I think I have correctly rendered the argument. It is to be noted that the parabola was at this period always thought of in the form we should now denote by the equation $y = x^2$, and the figure referred to by Leibniz is that which Wallis calls the complement of the semiparabola.

fore also a circle will be found equal to the surface of the parabolic conoid; but this is not the place to deduce it at full length.

Now these, which may seem to be great matters, are only the very simplest results to be obtained by this calculus; for many much more important consequences follow from it, nor does there occur any simple problem in geometry, either pure or applied to mechanics, that can altogether evade its power. Now we will expound the elements of the calculus itself.

The fundamental principle of the calculus.

Differences and sums are the inverses of one another, that is to say, the sum of the differences of a series is a term of the series, and the difference of the sums of a series is a term of the series; and I enunciate the former thus, $\int dx = x$, and the latter thus, $d \int x = x$.

Thus, let the differences of a series, the series itself, and the sums of the series, be, let us say,

Diffs.		1	2	3	4	5	....	dx
Series	0	1	3	6	10	15	....	x
Sums		0	1	4	10	20	25	..	$\int x$

Then the terms of the series are the sums of the differences, or $x = \int dx$; thus, $3 = 1 + 2$, $6 = 1 + 2 + 3$, etc.; on the other hand, the differences of the sums of the series are terms of the series, or $d \int x = x$; thus, 3 is the difference between 1 and 4, 6 between 4 and 10.

Also $da = 0$, if it is given that a is a constant quantity, since $a - a = 0$.

Addition and Subtraction.

The difference or sum of a series, of which the general term is made up of the general terms of other series by addition or subtraction, is made up in exactly the same manner from the differences or sums of these series; or

$$x + y - v = \int \overline{dx + dy - dv}, \quad \int \overline{x + y - v} = \int x + \int y - \int v.$$

This is evident at sight, if you take any three series, set out their sums and their differences, and take them together correspondingly as above.

Simple Multiplication.

Here $\qquad dxy = x\,dx + y\,dy$, or $xy = \int x\,dx + \int y\,dy$.

This is what we said above about figures taken together with their complements being equal to the circumscribed rectangle. It is demonstrated by the calculus as follows:

dxy is the same thing as the difference between two successive xy's; let one of these be xy, and the other $x + dx$ into $y + dy$; then we have

$$dxy = \overline{x + dx} \cdot \overline{y + dy} - xy = x\,dy + y\,dx + dx\,dy\,;$$

the omission of the quantity $dx\,dy$, which is infinitely small in comparison with the rest, for it is supposed that dx and dy are infinitely small (because the lines are understood to be continuously increasing or decreasing by very small increments throughout the series of terms), will leave $x\,dy + y\,dx$; the signs vary according as y and x increase together, or one increases as the other decreases; this point must be noted.

Simple Division.

Here we have $\qquad d\dfrac{y}{x} = \dfrac{x\,dy - y\,dx}{xx}$.

For, $d\dfrac{y}{x} = \dfrac{y + dy}{x + dx} - \dfrac{y}{x} = \dfrac{x\,dy - y\,dx}{xx + x\,dx}$, which becomes (if we write xx for $xx + x\,dx$, since $x\,dx$ can be omitted as being infinitely small in comparison with xx) equal to $\dfrac{x\,dy - y\,dx}{xx}$; also, if $y = aa$, then $dy = 0$, and the result becomes $-\dfrac{aadx}{xx}$, which is the value we used a little while before in the case of the tangent to the hyperbola.

From this any one can deduce by the calculus the rules for *Compound Multiplication and Division*; thus,

$$dxvy = xy\,dv + xv\,dy + yv\,dx,$$

$$d\frac{y}{vz} = \frac{vz\,dy - yv\,dz - yz\,dv}{vv.zz}\,;$$

as can be proved from what has gone before; for we have

$$d\frac{y}{x} = \frac{x\,dy - y\,dx}{xx}\,;$$

hence, putting zv for x, and $z\,dv + v\,dz$ for dx or dzv in the above, we obtain what was stated.

Powers follow: $dx^2 = 2x\,dx$, $dx^3 = 3x^2\,dx$, and so on. For, putting $y = x$, and $v = x$, we can write dx^2 for dxy, and this is (from above) equal to $x\,dy + y\,dx$, or (if $x = y$, and consequently $dx = dy$) equal to $2x\,dx$. Similarly, for dx^3 we write $dxyv$, that is (from above) $xy\,dv + xv\,dy + yv\,dx$, or (putting x for y and v and dx for dy and dv) equal to $3x^2dx$. Q. E. D. By the same method, in general, $dx^e = e.x^{e-1}\,dx$, as can easily be proved from what has been said.

Hence also,
$$d\,\frac{1}{x^h} = -\,\frac{h\,dx}{x^{h+1}}\,.$$

For, if $\dfrac{1}{x^h} = x^e$, then $e = -h$, and $x^{e-1} = \dfrac{1}{x^{h+1}}$, as is well known to any one who understands the nature of the exponents in a geometrical progression. The same thing will do for *fractions*. The procedure is the same for irrationals or *Roots*. $d.\sqrt[r]{x^h} = dx^{h:r}$, (where by $h:r$ I mean h/r, or h divided by r), or dx^e (taking e equal to h/r), or $e.x^{e-1}\,dx$, by what has been said above, or (by substituting once more $h:r$ for e, and $\overline{h-r}:r$ for $e-1$) $\dfrac{h}{r}\,.\,x^{\overline{h-r:r}}\,.dx$; and thus finally we get the value of $d.\sqrt[r]{x^h}$.

Moreover, conversely, we have
$$\int .x^e\,dx = \frac{x^{e+1}}{e+1}\,,\quad \int \frac{1}{x^e}\,dx = -\,\frac{1}{e-1.x^{e-1}}\,,\quad \int .\sqrt[r]{x^h}.dx = \frac{r}{r+h}\,\sqrt[r]{.x^{\overline{h+r:r}}}\,.$$

These are the elementary principles of the differential and summatory calculus, by means of which highly complicated formulas can be dealt with, not only for a fraction or an irrational quantity, or anything else; but also an indefinite quantity, such as x or y, or any other thing expressing generally the terms of any series, may enter into it.

§ 17.

The next manuscript bears no date; but this can be easily assigned to a certain extent, from internal evidence. It is for one thing later than the publication in the *Acta Eruditorum* of Leibniz's first communication to the world of his calculus in 1684. The manuscript is an answer, or rather the first rough draft probably of such an answer, to the animadversions of Bernhard Nieuwentijt against the idea of the infinitesimal calculus. The latter stated that (i) Leibniz could explain no more than Barrow or

Newton how the infinitely small differences differed from absolute zero; (ii) it was not clear how the differentials of higher order were obtained from those of the first order; (iii) the differential method cannot be applied to exponential functions. Leibniz answers the first point skilfully, fails over the second through erroneous work, which I think he afterward perceived; for he has a note that the whole thing is to be carefully revised before publication. It almost seems that he was not quite confident in his own powers of completely answering these objections, for he also notes that the rudeness of language in which the answer is commenced must be mollified.

On the third point he is silent; in the later written *Historia,* we have seen he is able to get, not over, but round the difficulty of the exponential function; but the silence here would seem to say that Leibniz could not manage exponentials as yet.

The success of the answer to the first point is due to the underlying principle that the ratio $dy: dx$ ultimately becomes a *rate*; when this idea is muddled by an admixture of the infinitesimal idea in the last paragraph the result is almost disastrous. Leibniz, however, looked on his calculus as a tried tool more than anything else.

When my infinitesimal calculus, which includes the calculus of differences and sums, had appeared and spread, certain over-precise veterans began to make trouble; just as once long ago the Sceptics opposed the Dogmatics, as is seen from the work of Empicurus against the mathematicians (i. e., the dogmatics), and such as Francisco Sanchez, the author of the book *Quod nihil scitur,* brought against Clavius; and his opponents to Cavalieri, and Thomas Hobbes to all geometers, and just lately such objections as are made against the quadrature of the parabola by Archimedes by that renowned man, Dethlevus Cluver. When then our method of infinitesimals, which had become known by the name of the calculus of differences, began to be spread abroad by several examples of its use, both of my own and also of the famous brothers Bernoulli, and more especially by the elegant writings of that illustrious Frenchman, the Marquis d'Hospital, just lately a certain erudite mathematician, writing under an assumed name in the scientific *Journal de Trevoux,* appeared to find fault with this method. But to mention one of them by name, even before this there arose against me in Holland Bernard Nieuwentijt, one indeed really well equipped both in learning and ability, but one who wished rather to become known by revising our methods to some extent than by advancing them. Since I introduced not only the first differences, but also the second, third and other higher differences, inassignable or incomparable with these first differences, he wished to appear satisfied with the first only; not considering that the same difficulties existed in the first as in the others that followed, nor that wherever they might be overcome in the first, they also ceased to appear in the rest. Not to mention how a very learned young man, Hermann of Basel, showed that the second and higher differences were avoided by the former in name only, and not in reality; moreover, in demonstrating theorems by the legitimate use of the first differ- ences, by adhering to which he might have accomplished some useful work on his own account, he fails to do so, being driven to fall back on assumptions that are admitted by no one; such as that something different is obtained by multiplying 2 by m and by multiplying m by 2; that the latter was impossible in any case in which the former was possible; also that the square or cube of a quantity is not a quantity or Zero.

In it, however, there is something that is worthy of all praise, in that he desires that the differential calculus should be strength- ened with demonstrations, so that it may satisfy the rigorists; and this work he would have procured from me already, and more willingly, if, from the fault-finding everywhere interspersed, the wish had not appeared foreign to the manner of those who desire the truth rather than fame and a name.

It has been proposed to me several times to confirm the essen- tials of our calculus by demonstrations, and here I have indicated below its fundamental principles, with the intent that any one who has the leisure may complete the work. Yet I have not seen up to the present any one who would do it. For what the learned Hermann has begun in his writings, published in my defence against Nieuwentiit, is not yet complete.

For I have, beside the mathematical infinitesimal calculus, a method also for use in Physics, of which an example was given in the *Nouvelles de la République des Lettres*; and both of these I include under the Law of Continuity; and adhering to this, I have shown that the rules of the renowned philosophers Descartes and Malebranche were sufficient in themselves to attack all problems on Motion.

I take for granted the following postulate:

In any supposed transition, ending in any terminus, it is permissible to institute a general reasoning, in which the final terminus may also be included.

For example, if A and B are any two quantities, of which the former is the greater and the latter is the less, and while B remains the same, it is supposed that A is continually diminished, until A becomes equal to B; then it will be permissible to include under a general reasoning the prior cases in which A was greater than B, and also the ultimate case in which the difference vanishes and A is equal to B. Similarly, if two bodies are in motion at the same time, and it is assumed that while the motion of B remains the same, the velocity of A is continually diminished until it vanishes altogether, or the speed of A becomes zero; it will be permissible to include this case with the case of the motion of B under one general reasoning. We do the same thing in geometry, when two

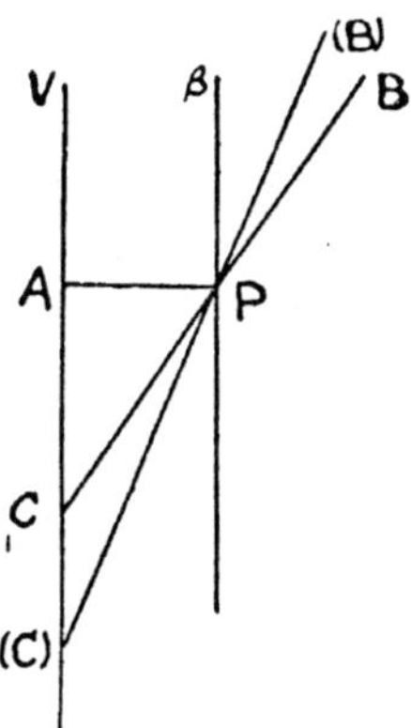

straight lines are taken, produced in any manner, one VA being given in position or remaining in the same site, the other BP passing through a given point P, and varying in position while the point P remains fixed; at first indeed converging toward the line VA and meeting it in the point C; then, as the angle of inclination VCA is continually diminished, meeting VA in some more remote point (C), until at length from BP, through the position (B)P, it comes to βP, in which the straight line no longer converges toward VA, but is parallel to it, and C is an impossible or imaginary point. With this supposition it is permissible to include under some one general reasoning not only all the intermediate cases such as (B)P but also the ultimate case βP.

Hence also it comes to pass that we include as one case ellipses and the parabola, just as if A is considered to be one focus of an ellipse (of which V is the given vertex), and this focus remains fixed, while the other focus is variable as we pass from ellipse to ellipse, until at length (in the case when the line BP, by its inter-section with the line VA, gives the variable focus) the focus C becomes evanescent[73] or impossible, in which case the ellipse passes into a parabola. Hence it is permissible with our postulate that a parabola should be considered with ellipses under a common rea-soning. Just as it is common practice to make use of this method in geometrical constructions, when they include under one general construction many different cases, noting that in a certain case the converging straight line passes into a parallel straight line, the angle between it and another straight line vanishing.

Moreover, from this postulate arise certain expressions which are generally used for the sake of convenience, but seem to con-tain an absurdity, although it is one that causes no hindrance, when its proper meaning is substituted. For instance, we speak of an imaginary point of intersection as if it were a real point, in the same manner as in algebra imaginary roots are considered as ac-cepted numbers. Hence, preserving the analogy, we say that, when the straight line BP ultimately becomes parallel to the straight line VA, even then it converges toward it or makes an angle with it, only that the angle is then infinitely small; similarly, when a body ultimately comes to rest, it is still said to have a velocity, but one that is infinitely small; and, when one straight line is equal to another, it is said to be unequal to it, but that the difference is infinitely small; and that a parabola is the ultimate form of an ellipse, in which the second focus is at an infinite distance from the given focus nearest to the given vertex, or in which the ratio of PA to AC, or the angle BCA, is infinitely small.

Of course it is really true that things which are absolutely equal have a difference which is absolutely nothing; and that straight lines which are parallel never meet, since the distance

---

[73] The term is here used with the idea of "vanishing into the far distance."

between them is everywhere the same exactly; that a parabola is not an ellipse at all, and so on. Yet, a state of transition may be imagined, or one of evanescence, in which indeed there has not yet arisen exact equality or rest or parallelism, but in which it is passing into such a state, that the difference is less than any assignable quantity; also that in this state there will still remain some difference, some velocity, some angle, but in each case one that is infinitely small; and the distance of the point of intersection, or the variable focus, from the fixed focus will be infinitely great, and the parabola may be included under the heading of an ellipse (and also in the some manner and by the same reasoning under the heading of a hyperbola), seeing that those things that are found to be true about a parabola of this kind are in no way different, for any construction, from those which can be stated by treating the parabola **rigorously.**

Truly it is very likely that Archimedes, and one who seems so have surpassed him, Conon, found out their wonderfully elegant theorems by the help of such ideas; these theorems they completed with *reductio ad absurdum* proofs, by which they at the same time provided rigorous demonstrations and also concealed their methods. Descartes very appropriately remarked in one of his writings that Archimedes used as it were a kind of metaphysical reasoning (Caramuel would call it metageometry), the method being scarcely used by any of the ancients (except those who dealt with quadratrices); in our time Cavalieri has revived the method of Archimedes, and afforded an opportunity for others to advance still further. Indeed Descartes himself did so, since at one time he imagined a circle to be a regular polygon with an infinite number of sides, and used the same idea in treating the cycloid; and Huygens too, in his work on the pendulum, since he was accustomed to confirm his theorems by rigorous demonstrations; yet at other times, in order to avoid too great prolixity, he made use of infinitesimals; as also quite lately did the renowned La Hire.

For the present, whether such a state of instantaneous transition from inequality to equality, from motion to rest, from convergence to parallelism, or anything of the sort, can be sustained in a rigorous or metaphysical sense, or whether infinite extensions successively greater and greater, or infinitely small ones successively **less and less, are legitimate** considerations, is a matter that I own to be possibly open to question; but for him who would discuss these matters, it is not necessary to fall back upon metaphysical controversies, such as the composition of the continuum, or to make geometrical matters depend thereon. Of course, there is no doubt that a line may be considered to be unlimited in any manner, and that, if it is unlimited on one side only, there can be added to it something that is limited on both sides. But whether a straight line of this kind is to be considered as one whole that can be referred to computation, or whether it can be allocated among quantities which may be used in reckoning, is quite another question that need not be discussed at this point.

It will be sufficient if, when we speak of infinitely great (or more strictly unlimited), or of infinitely small quantities (i. e., the very least of those within our knowledge), it is understood that we mean quantities that are indefinitely great or indefinitely small, i. e., as great as you please, or as small as you please, so that the error that any one may assign may be less than a certain assigned quantity. Also, since in general it will appear that, when any small error is assigned, it can be shown that it should be less, it follows that the error is absolutely nothing; an almost exactly similar kind of argument is used in different places by Euclid, Theodosius and others; and this seemed to them to be a wonderful thing, although it could not be denied that it was perfectly true that, from the very thing that was assumed as an error, it could be inferred that the error was non-existent. Thus, by infinitely great and infinitely small, we understand something indefinitely great, or something indefinitely small, so that each conducts itself as a sort of class, and not merely as the last thing of a class. If any one wishes to understand these as the ultimate things, or as truly infinite, it can be done, and that too without falling back upon a controversy about the reality of extensions, or of infinite continuums in general, or of the infinitely small, ay, even though he think that such things are utterly impossible; it will be sufficient simply to make use of them as a tool that has advantages for the purpose of the calculation, just as the algebraists retain imaginary roots with great profit. For they contain a handy means of reckoning, as can manifestly be verified in every case in a rigorous manner by the method already stated.

But it seems right to show this a little more clearly, in order that it may be confirmed that the algorithm, as it is called, of our differential calculus, set forth by me in the year 1684, is quite reasonable. First of all, the sense in which the phrase *"dy* is the element of *y,"* is to be taken will best be understood by considering a line AY referred to a straight line AX as axis.

Let the curve AY be a parabola, and let the tangent at the vertex A be taken as the axis. If AX is called *x,* and AY, *y,* and the latus-rectum is *a,* the equation to the parabola will be $xx = ay$, and this holds good at every point. Now, let $A_1X = x$, and $_1X_1Y = y$

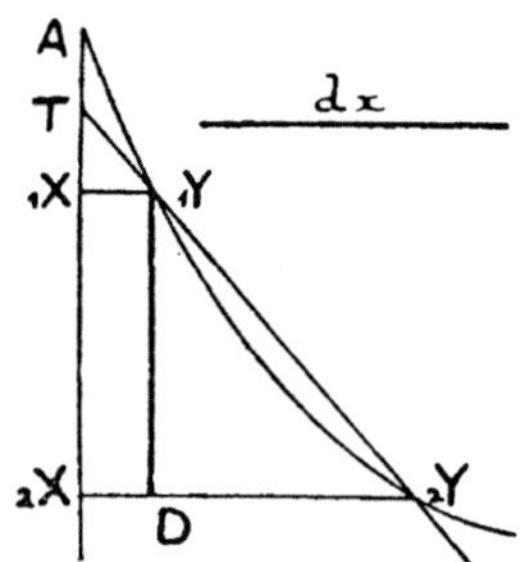

and from the point $_1$Y let fall a perpendicular $_1$YD to some greater ordinate $_2X_2Y$ that follows, and let $_1X_2X$, the difference between A_1X and A_2X, be called *dx*; and similarly, let D_2Y, the difference between $_1X_1Y$ and $_2X_2Y$, be called *dy.*

Then, since $y = xx : a$, by the same law, we have

$$y + dy = xx + 2x\,dx + dx\,dx, : a ;$$

and taking away the *y* from the one side and the $xx : a$ from the other, we have left

$$dy : dx = 2x + dx : a ;$$

and this is a general rule, expressing the ratio of the difference of the ordinates to the difference of the abscissae, or, if the chord $_1Y_2Y$ is produced until it meets the axis in T, then the ratio of the ordinate $_1X_1Y$ to T_1X, the part of the axis intercepted between the point of intersection and the ordinate, will be as $2x + dx$ to *a*. Now, since by our postulate it is permissible to include under the one general reasoning the case also in which the ordinate $_2X_2Y$ is moved up nearer and nearer to the fixed ordinate $_1X_1Y$ until it ultimately coincides with it, it is evident that in this case *dx* becomes equal to zero and should be neglected, and thus it is clear that, since in this case T_1Y is the tangent, $_1X_1Y$ is to T_1X as $2x$ is to *a*.

Hence, it may be seen that there is no need in the whole of our differential calculus to say that those things are equal which have a difference that is infinitely small, but that those things can be taken as equal that have not any difference at all, provided that the calculation is supposed to be general, including both the cases in which there is a difference and in which the difference is zero; and provided that the difference is not assumed to be zero until the calculation is purged as far as is possible by legitimate omissions, and reduced to ratios of non-evanescent quantities, and we finally come to the point where we apply our result to the ultimate case.

Similarly, if $x^3 = aay$, then we have

$$x^3 + 3xx\,dx + 3x\,dx\,dx + dx\,dx\,dx = aay + aa\,dy,$$

or cancelling from each side,

$$3xx\,dx + 3x\,dx\,dx + dx\,dx\,dx = aa\,dy,$$

or $\qquad 3xx + 3x\,dx + dx\,dx, : aa = dy : dx = {}_1X\,{}_1Y : T\,{}_1X;$

hence, when the difference vanishes, we have

$$3xx : aa = {}_1X\,{}_1Y : T\,{}_1X.$$

But if it is desired to retain dy and dx in the calculation, so that they may represent non-evanescent quantities even in the ultimate case, let any assignable straight line be taken as (dx), and let the straight line which bears to (dx) the ratio of y or ${}_1X\,{}_1Y$ to ${}_1XT$ be called (dy); in this way dy and dx will always be assignables bearing to one another the ratio of $D\,{}_2Y$ to $D\,{}_1Y$, which latter vanish in the ultimate case.

[Leibniz here gives a correction for a passage in the *Acta Eruditorum*, which is unintelligible without the context.]

On these suppositions, all the rules of our algorithm, as set out in the *Acta Eruditorum* for October 1684, can be proved without much trouble.

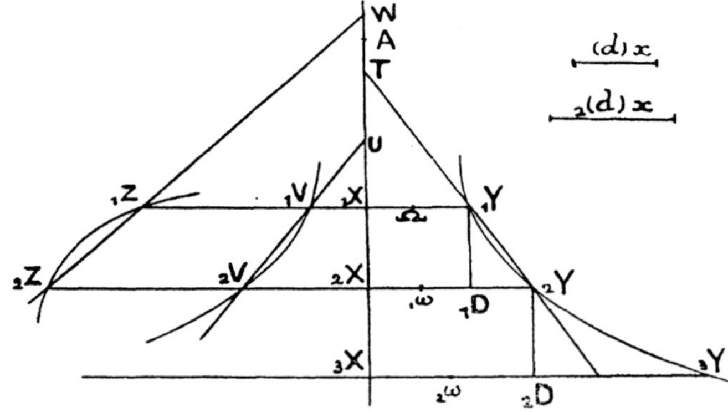

Let the curves YY, VV, ZZ be referred to the same axis AXX; and to the abscissae $A\,{}_1X\,(=x)$ and $A\,{}_2X\,(=x+dx)$ let there correspond the ordinates ${}_1X\,{}_1Y\,(=y)$ and ${}_2X\,{}_2Y\,(=y+dy)$, and also the ordinates ${}_1X\,{}_1V\,(=v)$ and ${}_2X\,{}_2V\,(=v+dv)$, and the ordinates

$_1$X $_1$Z ($=z$) and $_2$X $_2$Z ($=z+dz$). Let the chords $_1$Y $_2$Y, $_1$V $_2$V, $_1$Z $_2$Z, when produced meet the axis AXX in T, U, W. Take any straight line you will as $(d)x$, and, while the point $_1$X remains fixed and the point $_2$X approaches $_1$X in any manner, let this remain constant, and let $(d)y$ be another line which bears to $(d)x$ the ratio of y to $_1$XT, or of dy to dx; and similarly, let $(d)v$ be to $(d)x$ as v to $_1$XU or dv to dx; also let $(d)z$ be to $(d)x$ as z to $_1$XW or dz to dx; then $(d)x$, $(d)y$, $(d)z$, $(d)w$ will always be ordinary or assignable straight lines.

Nor for *Addition and Subtraction* we have the following:

$$\text{If } y-z=v, \text{ then } (d)y-(d)z=(d)v.$$

This I prove thus: $y+dy-z-dz=v+dv$, (if we suppose that as y increases, z and v also increase; otherwise for decreasing quantities, for z say, $-dz$ should be taken instead of dz, as I mentioned once before); hence, rejecting the equals, namely $y-z$ from one side, and v from the other, we have $dy-dz=dv$, and therefore also $dy-dz:dx=dv:dx$. But $dy:dx$, $dz:dx$, $dv:dx$ are respectively equal to $(d)y:(d)x$, $(d)z:(d)x$, and $(d)v:(d)x$. Similarly, $(d)z:(d)y$ and $(d)v:(d)y$ are respectively equal to $dz:dy$ and $dv:dy$. Hence, $(d)y-(d)z,:(d)x=(d)v:(d)x$; and thus $(d)y-(d)z$ is equal to $(d)v$, which was to be proved; or we may write the result as $(d)v:(d)y=1-(d)z:(d)y.$

This rule for addition and subtraction also comes out by the use of our postulate of a common calculation, when $_1$X coincides with $_2$X, and $_1$YT, $_1$YU, $_1$YW are the tangents to the curves YY, VV, ZZ. Moreover, although we may be content with the assignable quantities $(d)y$, $(d)v$, $(d)z$, $(d)x$, etc., since in this way we may perceive the whole fruit of our calculus, namely a construction by means of assignable quantities, yet it is plain from what I have said that, at least in our minds, the unassignables dx and dy may be substituted for them by a method of supposition even in the case when they are evanescent; for the ratio $dy:dx$ can always be reduced to the ratio $(d)y:(d)x$, a ratio between quantities that are assignable or undoubtedly real. Thus we have in the case of tangents $dv:dy=1-dz:dx$, or $dv=dy-dz$.

Multiplication. Let $ay=xv$, then $a(d)y=x(d)v+v(d)x.$
Proof. $ay+a\,dy=x+dx,\ v+dv=xv+x\,dv+v\,dx+dx\,dv$;
and, rejecting the equals ay and xy from the two sides,

$$a\,dy = x\,dv + v\,dx + dx\,dv,$$

or

$$\frac{a\,dy}{dx} = \frac{x\,dv}{dx} + v + dv;$$

and transferring the matter, as we may, to straight lines that never become evanescent, we have

$$\frac{a(d)y}{(d)x} + \frac{x(d)y}{(d)x} + v + dv;$$

so that, since it alone can become evanescent, dv is superfluous, and in the case of the vanishing differences, as in that case $dv=0$, we have

$$a(d)y = x(d)v + v(d)x, \text{ as was stated,}$$

or $\qquad (d)y : (d)x = x + v, : a.$

Also, since $(d)y :(d)x$ always $= dy: dx$, it will be allowable to suppose this is true in the case when dy, dx become evanescent, and to say that $dy: dx = x + v: a$, or $a\,dy = x\,dv + v\,dx$.

Division.   Let $z:a = v:x$, then $(d)z:a = v(d)x - x(d)y, :xx$.

Proof $\qquad\qquad z + dz : a = v + dv, : ,x + dx;$

or clearing of fractions, $xz + xdz + zdx + dzdx = av + adv$; taking away the equals xz and av from the two sides, and dividing what is left by dx, we have

$$a\,dv - x\,dz, : dx = z + dz,$$

$$\text{or } a(d)v - x(d)z, : dx = z + dz;$$

and thus, only dz, which can become evanescent, is superfluous. Also, in the case of vanishing differences, when $_1$X coincides with $_2$X, since in that case $dz=0$, we have

$$a(d)v - x(d)z, :(d)x = z = av : x;$$

whence, (as was stated) $(d)z = ax(d)v - av(d)x, : xx,$

or $\qquad (d)z: (d)x = (a:x)(d)v: (d)x - av: xx.$

Also, since $(d)z:(d)x$ is always equal to $dz:dx$, on all other occasions, it is allowable to suppose this to be so also when dz, dv, dx are evanescent, and to put

$$dz: dx = ax\,dv - av\,dx, :xx$$

For *Powers,* let the equation be $a^{\underline{n-e}}x^e = y^n$, then

$$\frac{(d)\,y}{(d)\,x} = \frac{e.x^{\underline{e-1}}}{n.y^{\underline{n-1}}};$$

and this I will prove in a manner a little more detailed than those above, thus:

$$a^{n-e}, \frac{1}{1}x^{e} + \frac{e}{1}x^{e-1}\,dx + \frac{e,e-1}{1,2}\,x^{e-2}\,dxdx + \frac{e,e-1,e-2}{1,2,3}\,x^{e-3}\,dxdxdx$$

(and so on until the factor $e-e$ or 0 is reached)

$$= \frac{1}{1}y^{n} + \frac{n}{1}y^{n-1}\,dy + \frac{n,n-1}{1,2}\,y^{n-2}\,dydy + \frac{n,n-1,n-2}{1,2,3}\,y^{n-3}\,dydydy$$

(and so on until the factor $n-n$ or 0 is reached);

take away from the one side $a^{n-e}x^{e}$, and from the other side y^{n}, these being equal to one another, and divide what is left by dx, and lastly, instead of the ratio $dy:dx$, between the two quantities that continually diminish, substitute the ratio that is equal to it, $(d)y:(d)x$, a ratio between two quantities, of which one, $(d)x$, always remains the same during the time that the differences are diminishing, or while $_2$X is approaching the fixed point $_1$X and we have

$$\frac{e}{1}x^{e-1} + \frac{e,e-1}{1,2}\,x^{e-2}\,dx + \frac{e,e-1,e-2}{1,2,3}\,x^{e-3}\,dxdx + \text{etc.}$$

$$= \frac{n}{1}y^{n-1}\frac{(d)y}{(d)x} + \frac{n,n-1}{1,2}\,y^{n-2}\frac{(d)y}{(d)x}\,dy + \frac{n,n-1,n-2}{1,2,3}\,y^{n-3}\frac{(d)y}{(d)x}\,dydy + \text{etc.}$$

Now, since by the postulate there is included in this general rule the case also in which the differences become equal to zero, that is when the points $_2$X, $_2$Y coincide with the points $_1$X, $_1$Y respectively; therefore, in that case, putting dx and dy equal to 0, we have

$$\frac{e}{1}x^{e-1} = \frac{n}{1}y^{n-1}\frac{(d)y}{(d)x},$$

the remaining terms vanishing, or $(d)y : (d)x = e.x^{x-1} : n.y^{n-1}$. Moreover, as we have explained, the ratio $(d)y:(d)x$ is the same as the ratio of y, or the ordinate $_1$X$_1$Y, to the subtangent $_1$XT, where it is supposed that T$_1$Y touches the curve in $_1$Y.

This proof holds good whether the powers are integral powers or roots of which the exponents are fractions. Though we may also get rid of fractional exponents by raising each side of the equation to some power, so that e and n will then signify nothing else but powers with rational exponents, and there will be no need of a series proceeding to infinity. Moreover, at any rate, it will be permissible, by means of the explanation given above, to return to the unassignable quantities dy and dx, by making in the case of evanescent differences, as in all other cases, the supposition that the ratio of the evanescent quantities dy and dx is equal to the ratio of $(d)y$ and $(d)x$, because this supposition can always be reduced to an undoubtable truth.

Thus far the algorithm has been demonstrated for differences of the first order: now I will proceed to show that the same method will hold good for the differences of the differences. For this purpose, take three ordinates, $_1X\,_1Y$, $_2X\,_2Y$, $_3X\,_3Y$, of which $_1X\,_1Y$ remains constant, but $_2X\,_2Y$ and $_3X\,_3Y$ continually approach $_1X\,_1Y$ until finally they both coincide with it simultaneously; which will happen if the speed with which $_3X$ approaches $_1X$ is to the speed with which $_2X$ approaches $_1X$ is in the ratio of $_1X\,_3X$ to $_1X\,_2X$. Also let two straight lines be assigned, $(d)x$ always constant for any position of $_2X$, and $_2(d)x$ for any position of $_3X$; also let $(d)y$ always be to $(d)x$ as D_2Y is to $_1X\,_2X$, or as y (i. e., $_1X\,_1Y$) is to $_1XT$; thus, while $(d)x$ remains always the same, $(d)y$ will be altered as $_2X$ approaches $_1X$; similarly, let $_2(d)y$ be to $_2(d)x$ as $_2D_3Y$ to $_2X\,_3X$ or as $y+dy$ (i. e., $_2X\,_2Y$) to $_2X\,_2T$; thus while $_2(d)x$ remains constant, $_2(d)y$ will be altered as $_3X$ approaches $_1X$.

Also let $(d)y$ be always taken in the varying line $_2X\,_2Y$, and let $_2X\,_1\omega$ be equal to $(d)y$, and similarly take $_2(d)y$ in the line $_3X\,_3Y$, and let $_3X\,_2\omega$ be equal to $_2(d)y$. Thus, while $_2X$ and $_3X$ continually approach to the straight line $_1X\,_1Y$, $_2X\,_1\omega$ and $_3X\,_2\omega$ continually approach it also, and finally coincide with it at the same time as

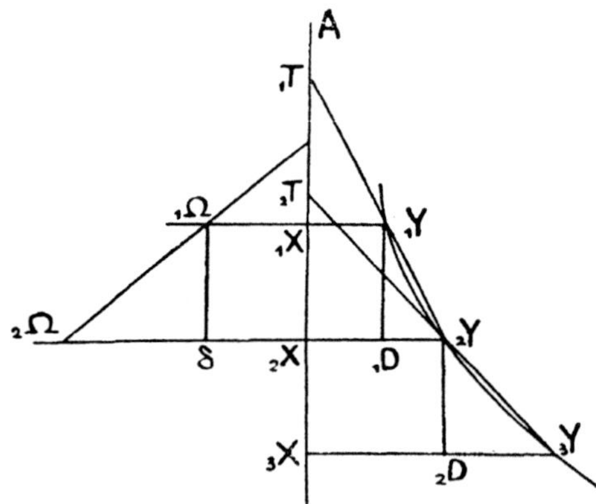

$_2X$ and $_3X$. Further, let the point in the ordinate $_1X\,_1Y$, which $_1\omega$ continually approaches and with which it at last coincides, be marked, and let it be Ω; then $_1X\Omega$ is the ultimate $(d)y$, which bears to $(d)x$ the ratio of the ordinate $_1X\,_1Y$ to the subtangent $_1XT$, where it is supposed that $T\,_1X$ touches the curve in $_1Y$, because then indeed $_1Y$ and $_2Y$ coincide. Now, since all this can be done, no matter where $_1Y$ may be taken on the curve, it is evident that a curve $\Omega\Omega$ will be produced in this way, which is the differentrix of the curve YY; just as, conversely, the curve YY is the summatrix curve of $\Omega\Omega$, as can be readily demonstrated.

By this method, the calculus may be demonstrated also for the differences of the differences.

Let $_1X\,_1Y$, $_2X\,_2Y$, $_3X\,_3Y$ be three ordinates, of which the values are y, $y+dy$, $y+dy+ddy$, and let $_1X\,_2X\,(dx)$ and $_2X\,_3X\,(dx+ddx)$ be any distances, and $D\,_2Y\,(dy)$ and $_2D\,_3Y\,(dy+ddy)$ the differences. Now the difference between $(d)y$ and $_2(d)y$, or between $_1X\Omega$ and $_2X\,_2\Omega$ is $\delta\,_2\Omega$, and that between $_1X\,_2X$ and $_2X\,_3X$ is ddx; also let

$$(d)\,dx:(d)\,x=dx:\,_2(d)\,x,\quad ^{74}\quad \text{and similarly let}$$

$$(d)\,dy:(d)\,y=\,_2\Omega\delta:\,_1X\,_2X\ \text{ or }\ _1X\Omega:\,_1XT.$$

Now, for the sake of example, let us take $ay=xv$. Then we have $a\,dy=x\,dv+v\,dx+dx\,dv$, as has been shown above; and similarly,

$$a\,dy+a\,ddy=(x+dx)(dv+ddv)+(v+dv)(dx+ddx)\qquad ^{75}$$
$$+(dx+ddx)(dv+ddv)$$
$$=x\,dv+x\,ddv+dx\,dv+dx\,ddv+v\,dx+v\,ddx$$
$$+\,dv\,dx+dv\,ddx+dx\,dv+dx\,ddv$$
$$+\,ddx\,dv+ddx\,ddv.$$

Taking away $a\,dy$ from one side, and $x\,dx+v\,dx+dx\,dv$ from the other, there will be left in any case

$$\frac{ddy}{ddx}=\frac{ddy}{a\,ddx}+\frac{v}{a}+\frac{2\,dx\,dv}{a\,ddx}+\frac{2\,dv}{a}+\frac{2\,dx\,ddx}{a\,ddx}+\frac{ddv}{a}.$$

In this it is evident that the ratio between ddy and ddx can be expressed by the ratio of the straight line $(d)\,dy$ to $(d)\,x$, the straight line assumed above, which we have supposed to remain constant as $_2X$ and $_3X$ approach $_1X$. Also, since $(d)\,dx$, (since it bears an assignable ratio to $(d)\,x$, however nearly $_2X$ approaches to $_1X$, or

---

^{74} This makes $(d)\,dx$ an inassignable. It may be a misprint due to a slip of Leibniz, or of Gerhardt in transcription; for there is no similarity between it and the statement in the next line. I cannot however offer any feasible suggestion for correction.

^{75} This is quite wrong. Leibniz has evidently substituted $x+dx$ for x, etc.; which is not legitimate unless $_3X_3Y$ is taken as $y+dy+d(y+dy)$, and so on; even then fresh difficulties would be introduced. As it stands, this line should read

$$a\,dy+a\,ddy=x(dv+ddv)+v(dx+ddx)+(dx+ddx)(dv+ddv).$$

On account of this error and that noted above, there is not much profit in considering the remainder of this passage.

however much dx, the difference between the abscissae, is diminished), is not evanescent, even when, finally, dx and ddx, dv and ddv, are all supposed to be zero. In the same way, the ratio of ddv to ddx may be expressed by the ratio of an assignable straight line $(d)dv$ to the assumed constant $(d)x$; and even the ratio of $dv\,dx$ to $a\,ddx$ may be so expressed; for, since $dv:dx=(d)v:(d)x$, therefore $dv\,dx:dx\,dx=(d)v:(d)x$. Hence, if a new straight line, $(dd)x$, is assumed to be such that $a\,ddx:dx\,dx=(dd)x:(d)x$, then the new straight line will be assignable, even though dx, ddx, etc. become evanescent. Since therefore $dv\,dx:dx\,dx=(d)v:(d)x$ and $dv\,dx:a\,ddx=(d)x:(dd)x$, it follows that $dv\,dx:a\,ddx=(d)v:(dd)x$, and thus at length there is produced an equation that is freed as far as possible from those ratios that might become evanescent, namely,

$$\frac{(d)\,dy}{(d)dx}=\frac{x}{a}\frac{(d)\,dy}{(d)dx}+\frac{y}{a}+\frac{2\,(d)y}{(dd)x}+\frac{2\,dv}{a}+\frac{2\,dx}{a}\frac{(d)\,dy}{(d)dx}+\frac{ddv}{a}.$$

Thus far all the straight lines have been considered to be assignable so long as $_1X$ and $_2X$ do not coincide; but in the case of coincidence, dv and ddv are zero, and we have

$$\frac{(d)\,dy}{(d)dx}=\frac{x}{a}\frac{(d)\,dv}{(d)dx}+\frac{v}{a}+\frac{2\,(d)y}{(dd)x}+\frac{0}{a}+\frac{2\,(d)dv}{(d)dx}\frac{0}{a}+\frac{0}{a},$$

or, omitting terms equal to zero,

$$\frac{(d)\,dy}{(d)dx}=\frac{x}{a}\frac{(d)\,dv}{(d)dx}+\frac{v}{a}+\frac{2\,(d)y}{(dd)x}.$$

Hence, if dx, ddx, dv, ddv, dy, ddy, are by a certain fiction imagined to remain, even when they become evanescent, as if they were infinitely small quantities (and in this there is no danger, since the whole matter can be always referred back to assignable quantities), then we have in the case of coincidence of the point $_1X$ and $_2X$ the equation

$$\frac{ddv}{ddx}=\frac{x}{a}\frac{ddy}{ddx}+\frac{v}{a}+\frac{2}{a}\frac{dx\,dy}{ddx}.$$

VI.

LEIBNIZ IN LONDON.[1]

(BY C. I. GERHARDT.)

LEIBNIZ paid two visits to London from Paris, where he was staying from March, 1672, to October, 1676: from the Elector of Mainz, was from January 11 to the beginning of March, 1673; the second was made on his way home to Germany, when he stopped in London for about a week in October, 1676.

Leibniz had a habit of writing out all the important scientific points in the correspondence that he kept up with noted people, so that he might thus impress them the more deeply upon his memory. I have discovered among his manuscripts three folio sheets on which he has written down the things worth noting in connection with these two visits to London.[2] The sheets which relate to his second visit have been known to me for some time; but the other ones, referring to the first visit, I came across only during my last stay in Hanover in the summer vacation of the year 1890.

In what follows, I have only paid attention to the contents of these sheets which refer to mathematics.[3]

[1] Translated from an article by Dr. Gerhardt in the *Sitzungsberichte der Königlich Preussischen Akademie der Wissenschaften zu Berlin,* 1891, pp. 157-165, and published in *The Monist* for Oct., 1917. The notes are mine.

[2] These highly important documents ought to be photographed and published in facsimile.

[3] It seems a pity that Gerhardt has not given the contents of the section labeled "Mechanica," unless indeed this is all non-mathematical; there may

The sheet relating to Leibniz's first visit to London, of which I have added a partial transcript under the heading I, is divided on both pages into sections [the word used in the original is *Felder* = columns, but it will be seen that, according to the transcript given later, the sections are horizontal and not vertical], in which Leibniz has entered all that he considered to be worth noting. While the sections labeled "Chymica," "Mechanica," "Magnetica," "Botanica," "Anatomica," "Medica," and "Miscellanea" are filled up with an extraordinary number of memoranda, the first sections, which are allotted to mathematical subjects, are very poorly filled. That labeled "Geometrica" contains a note that is especially worth remarking: "Tangents to figures of all kinds. Development of geometrical figures by the motion of a point in a moving straight line."[4] In all probability it may be supposed that this refers to the lectures of Barrow, delivered on his method of tangents at the University of Cambridge down to the year 1669. As is well known, the method of Barrow is only applicable to such curves as can be expressed by rational functions.[5] Newton's name was mentioned in

be in it some intimation that would lead to a clue as to the origin of Leibniz's use of the word *moment*, meaning thereby, not Newton's use of the word, but the idea now familiar to us in the determination of the center of gravity of an area, expressed by the equation

$$x = \Sigma ax / \Sigma a,$$

where a is the element of the area distant x from the axis, x the distance of the center of gravity from that axis, and Σax is the sum of the 'first moments of the elements' or 'the first moment of the whole area.' See Note 18, below.

[4] *"Tangentes omnium figurarum. Figurarum geometricarum explicatio per motum puncti in moto lati."*

[5] In a footnote, Gerhardt asserts that "Barrow's *Lectiones Geometricae* appeared in 1672." This is incorrect; for they were published, combined with the second edition of the *Lectiones Opticæ*, in 1670; nor can Gerhardt be referring to the second edition, for that appeared in 1674 and then as a separate volume. Also, I have, in the little book on *The Geometrical Lectures of Isaac Barrow,* published by the Open Court Publishing Co., given reasons for supposing that these lectures were never delivered as *Lucasian Lectures,* though they may have formed the subject-matter for college lectures at Gresham and Trinity. Again, it is not true, although "well known," that "the method of Barrow was only applicable to such curves as can be expressed by rational functions"; this remark is even only partially true about the differential triangle method; for, as I have shown in the above-mentioned book, Barrow had a complete calculus, which included, among other things, the important idea of *substitu-*

the "Optica." Leibniz has the remark: "They told me about a certain phenomenon that Barrow confessed he was unable to solve. Newton's difficulty has so far not been solved, Father Pardies having given it up."[6] Obviously this remark applies to Newton's experiment on the refraction of light by a prism and to the decomposition of white sunlight, and especially to the fact that a circular solar image becomes after refraction a long spectrum. Father Pardies of Clermont had published in opposition to Newton his "Two Letters containing Animadversions upon I. Newton's Theory of Light," in the *Philosophical Transactions* of 1672, together with a letter from Newton.

It cannot be said for certain that Leibniz, during his first stay in London, met with any of the great English mathematicians; Wallis lived at Oxford, while Barrow and Newton resided at Cambridge.[7] Indeed, it is made a matter of plaint by Brewster, the biographer of Newton, that the Royal Society of London at that time numbered few men of distinguished talents who were in a position to perceive the truth of the optical discoveries of Newton. In the letter which Leibniz addressed to Oldenburg, the Secretary of the Royal Society, during his visit to London, he men-

tion, which is all that is necessary to complete the "*a*-and-*e*" method and make it applicable to surds and fractions, and probably was thus applied by Barrow in working out his constructions; but the whole thing was geometrical, which apparently hid the inner meaning until recently.

To my mind, the mention of but "tangents and local motion" points out that, on Leibniz's *first* reading of Barrow, he only perused at all carefully the first five lectures, which are relatively unimportant; or rather it confirms an opinion I had already expressed to Mr. P. E. B. Jourdain: see Note 43, p. 218.

[6] "*Locuti sunt mihi de phaenomeno quodam quod Barrovius fatetur se solvere non posse. Newtoni difficultas soluta hactenus non est, P. Pardies manus dante.*"

[7] It seems however that Leibniz attended the meetings of the Royal Society; at any rate once, when he exhibited the model of his calculating machine. It would be interesting if the roll of members present on all occasions during this period could be obtained, as doubtless they were kept. For such men as Ward were members at the time and attended the meetings, and Ward was, if not in the same class as the three whose names are given, an excellent mathematician; and, Leibniz, being somewhat of a notable, on account of his connection with the Embassy from Mainz, would surely be introduced to all eminent members present.

tioned that he had met by accident the mathematician Pell at the house of Boyle, the chemist. The conversation fell upon those number-series which in elementary mathematics were called the higher arithmetical series and whose sums and terms were found by the help of differences. Leibniz showed that he had gone deeply into the study of such series and had partly found out some new methods for calculating the terms.[8] Leibniz's letter to Oldenburg was dated Feb. 3, 1673 (1672 O. S.).[9]

From the preceding it appears that what Leibniz learned with reference to mathematics from his first visit to London was quite unimportant.[10] The chief aim of his stay in London was to be elected as a Fellow of the Royal Society; and this came to pass, owing in part to an exhibition of a model of his calculating machine, and in part to the friendly offices of Oldenburg.

After his return to Paris at the beginning of March, 1673, Leibniz was able to find more leisure to follow up his studies without hindrance; the political mission which was the cause of his being sent to Paris, was now at an end.

It may be regarded as certain that, before his first visit to London, Leibniz made the personal acquaintance of the men with whom he corresponded before he came to Paris, and especially Antoine Arnauld and de Carcavi. The

---

[8] The account given by Leibniz himself in the *Historia* (see above, Chapter III, p. 36) reads thus: "He" [for Leibniz wrote in the third person, under the guise of "a friend who knew all about the matter"] "also came across Pell accidentally, and described to him certain of his own observations on numbers, and the latter stated that they were not new, but it had been recently made known by Nikolaus Mercator.... This made Leibniz get the work of Nikolaus Mercator." As a matter of fact the suggested plagiarism, or what Leibniz took for such a suggestion, was from Mouton and not from Mercator. This is an instance of the lack of memory from which Leibniz suffered; such lack as caused him to make notes of all important points.

[9] See Note 32, p. 171, on the introduction of the Gregorian calendar.

[10] I cannot see what reason Gerhardt has for this statement, considering the contents of Barrow's book, which we know that Leibniz had purchased; that is, unless we assume either that Leibniz, as I have suggested, did not at that time read the whole of Barrow, or failed to grasp what Barrow had given owing to his (Leibniz's) incomplete knowledge of geometry.

latter belonged to the circle in which Pascal moved. Whether at that time Leibniz had made the acquaintance of Huygens is not quite so certain; at any rate he did not come into close relations with him until after his return from London. Huygens presented him with a copy of his great work, *Horologium Oscillatorium,* which had just (1673) been published. The recognition that his mathematical knowledge at that time was insufficient to enable him to understand the contents of this book, combined with a reawakening of his former love for mathematics, had the effect of making Leibniz devote himself with the greatest fervor to the study of mathematical subjects. Cavalieri's method of indivisible magnitudes, the writings of Gregory St. Vincent, the letters of Pascal (which were especially recommended to him by Huygens), were used by him as guides in his studies. As the first-fruits of these studies, he obtained the theorem that, when the square on the diameter of a circle was taken as unity, the area of the circle was expressed by the infinite series

$$1-\frac{1}{3}+\frac{1}{5}-\frac{1}{7}+\ \ldots\ldots\ldots\ \text{ad inf.} \qquad (11)$$

He obtained it thus: Instead of dividing the circle, as in the method of Cavalieri, into trapezia by means of parallels, he divided it into triangles by lines radiating from a point; the areas of these triangles being proportional to certain lines. With these lines as perpendicular ordinates a curve could be constructed that was divided by these ordinates into trapezia, each of which is double the corresponding triangle. In this way Leibniz obtained a curvilinear figure[12] whose area was double that of the circle, but which was expressed by a rational function, $x = y^2/(1 + y^2)$,[13]

[11] Leibniz's own date for the discovery of this result, usually alluded to by him as the "Arithmetical Tetragonism," is 1674; "But in the year 1674 (so much it is possible to state definitely) he came upon the well-known Arithmetical Tetragonism;...." (see above, Chap. III, p. 42).

[12] See the first critical note, pp. 172ff.

[13] See the first critical note, pp. 172ff.

of its coordinates; and, using a method that was similar to that employed by Mercator for the equilateral hyperbola, this area could be found (Quadratrix).[14]

For the rest of Leibniz's treatment, see the hitherto unpublished manuscript, given under II in the appendix that follows.

As was often the case in the first scientific studies of Leibniz, intimations of the great problems that occupied his attention his whole life through are found here in his first efforts in the domain of higher mathematics. First is it to be remarked that Leibniz abandoned the division of curvilinear figures into trapezia, as employed by Cavalieri, and instead divided them into triangles; from this he was led to the "characteristic triangle,"[15] which formed the foundation in the application of the differential calculus. Further, Leibniz constructed, instead of the proposed curve, another of which the area could be found (the "quadratrix" as he called it); this method of procedure frequently occurs in the later works of Leibniz on the integral calculus. Closely connected also with this is the solution of the inverse method of tangents, that is, given the tangent, to find the curve.

In these first efforts of Leibniz in the domain of higher mathematics is clearly to be seen the influence of his study of the writings of Pascal.[16]   The French mathematicians Roberval and Pascal did not consider that Cavalieri's

[14] Observe that Leibniz (or Gerhardt) employs this word in a different sense from that of Barrow, with whom it means the special curve whose equation is $y = (r - x)\tan \pi x/2r$, a curve that is particularly connected with the circle.

[15] This contradicts both Gerhardt and Leibniz himself, who said that he got it from a consideration of a figure used by Pascal in finding the content of the sphere.   See also the first critical note, pp. 172ff.

[16] I will consider this influence in connection with an essay by Gerhardt on this very point in the following chapter, when I shall endeavor to substantiate an opinion I have formed with regard to the earlier manuscripts of Leibniz, which were discovered by Gerhardt, and of which translations are given above, on pp. 59-114. I suggest that these do not represent so much the record of his original investigations as notes made while using the works of his predecessors as text-books.

method was consistent with the rigorous requirements of mathematics;[17] they reverted to the study of the Greek mathematicians, and especially to the writings of Archimedes, combining with their method the developments which Kepler, in particular, had brought about by the introduction of infinitely small magnitudes into geometry. Moreover, in connection with Pascal, it is to be observed that he generalized into a "barycentric calculus" the procedure used by Archimedes for the quadrature of the parabola by means of the equilibrium of the lever.[18] This "calculus" enabled him to solve problems on the cycloid which his contemporaries had vainly attempted.[19] It was not unknown to Leibniz that, since the time of Pappus of Alexandria, quadratures and cubatures had been calculated by the aid of the center of gravity (Guldin's rule, "Centrobaryca"); certainly he was now led, by the works of Pascal, again to notice the methods for the determination of the center of gravity, and was also induced to attempt to extend and perfect them. The manuscript of

[17] I fail to see how this statement can be completely reconciled with the following well-known quotation from the *"Lettre de A. Dettonville à Carcavy"* (1658):

"J'ay voulu faire cét advertissement pour monstrer que tout ce qui est demonstré par les veritables regles des indivisibles se demonstrera aussi à la rigueur et à la maniere des anciens; et qu'ainsi l'une de ces Methodes ne differe de l'autre qu'en la maniere de parler; ce qui ne peut blesser les personnes raissonnables quand on les a une fois avertyes de ce qu'on entend par là" (Vol. VIII, p. 352).

Pascal also says on p. 350: "*.... la doctrine des indivisibles, laquelle ne peut estre rejettée par ceux qui pretendent avoir rang entre les Geometres.*"

That is, the method of indivisibles does not differ from the method of exhaustions, except in the way the argument is put; and that the former must be accepted by any mathematician with pretensions to rank among geometers.

The page reference is to the edition of Pascal's Works in 14 volumes, in the series, *Les Grands Ecrivains de la France* (pub. Hachette et Cie., Paris, 1914).

[18] Pascal calls it "la balance." It is worth noting in this connection that Pascal uses the word *"force"* and not *"moment"* for the product of one of his weights and its lever-arm; so that we must look elsewhere for the clue to the use of the word "moment" in this sense by Leibniz.

[19] Several of the problems proposed were solved by Huygens, de Sluse, and Wren; but by special methods, which did not satisfy Pascal, who called for a general method. Later (1670) Barrow gives the rectification of the arc, as a special case of a general theorem (Lect. XII, App. 3, Ex. 2, see my *Barrow*, p. 177).

Leibniz which is dated October 25, October 26, October 29, November 1, 1675, and which contains the investigation on the center of gravity, is headed, "*Analysis Tetragonistica ex Centrobarycis.*"[20]

It is worth remarking that in this Leibniz continues the method by which he had found the series for the area of the circle. Incidentally these studies were the first occasion for the introduction of the symbol for a sum, i. e., the integral sign (October 29, 1675); from this as the antithesis, the sign for the difference, i. e., the symbol for differentiation, resulted.[21] The equation in which Leibniz first introduced the sign of integration was, in the notation of that time:

$$\frac{\overline{\text{omn. } l}^{\,\boxed{2}}}{2} \;\sqcap\; \text{omn. omn. } \overline{\frac{l}{a}}$$

that is,

$$\frac{(\text{omn.}l)^2}{2} = \text{omn. omn. } \overline{\frac{l}{a}} \; ;$$

for which Leibniz writes

$$\frac{\int \overline{l}^{\,2}}{2} \;\sqcap\; \int \overline{\int \overline{l}.\frac{l}{a}}$$

that is, when $l = dy$,

$$\frac{y^2}{2} = \frac{1}{a}\int dy \int dy$$

After his return to Paris in March, 1673, Leibniz was in constant communication with Oldenburg, the Secretary of the Royal Society; the subjects being almost entirely mathematical. In this way he obtained his knowledge of the work of the English mathematicians. Oldenburg's mentor on all mathematical questions was John Collins, who possessed a very wide acquaintance among English

---

[20] See pp. 65ff.

[21] See the second critical note, p. 179.

mathematicians; and it was through him that what they had done was communicated. In this respect special mention is to be made of the letter from Oldenburg to Leibniz, dated July 26, 1676, in which Collins informed him of a collection of letters from English mathematicians that he had in his possession. Collins mentions in it particularly that script of Newton, of December 10, 1672, in which the latter makes a communication about his method for tangents to curves, which are given by an explicit algebraical equation; he remarks that the method is only a corollary to a general procedure for solving other problems, such as those relating to rectification, determination of centers of gravity and so on.[22] Collins stated in addition that, besides what this letter showed, nothing further was known at that time about Newton's method. It was on account of these communications, and probably also on account of a letter from Newton to Oldenburg, of which Oldenburg sent a copy to Leibniz at Paris, that Leibniz was moved to make his return journey to Germany in October, 1676, by way of London. Leibniz stayed there about a week; he made the acquaintance of Collins, who willingly let him have access to his collection of treatises and letters.[23] What Leibniz found in them that he thought worth noting he set down

[22] Leibniz, in the *Acta Eruditorum* for the year 1700, says, "I can affirm that, when in 1684 I published the elements of my Calculus, I did not know any thing more of Mr. Newton's inventions in this kind, than what he formerly signified to me by his letters, viz., that he could find tangents without taking away surds;...." As Newton says in the article in *Phil. Trans.*, Vol. XXIX, No. 342, Anno 1714 (usually called the *Recensio*) this "is very extraordinary, and wants an explanation."

[23] This is feasible, but there is another alternative given by Dr. H. Sloman (*The Claim of Leibniz to the Invention of the Differential Calculus, English edition*, pub. Macmillan, 1860), which strikes me as even more probable. Sloman's points are as follows: (1) It is highly probable that Leibniz's week in London was the *last week of that month*. (2) Oldenburg had then in his possession two letters from Newton for Leibniz, dated Oct. 24 and 26; these he showed to Leibniz. (3) As Newton himself mentions, these were blotted and hastily written; and thus Leibniz asks, on this account, that Oldenburg should let him see the tract of Newton to which they refer; which tract Leibniz knew was in the possession of Oldenburg, that is, a copy of it. For the details of the argument, occupying ten quarto pages, see the above-mentioned book by Sloman, pp. 97-106.

on two folios; the one has the heading, *"Excerpta ex trac-
tatu Newtoni de Analysi per aequationes numero termi-
norum infinitas."* This is the paper which Newton sent in
June, 1699, to Barrow, from whom Collins received it on
July 30, 1699. Collins made a copy of it, and sent the
original back; and the original was printed in the year
1711. The other sheet has the heading, *"Excerpta ex Com-
mercio Epistolico inter Collinium et Gregorium."* A partial
transcript of both these sheets follows under the heading III.

With regard to the extracts from Newton's paper, it
is to be remarked that Leibniz was interested in the treatment of algebraical expressions of powers and in the turn-
ing of irrational expressions into the form of series by
means of division and root-extraction. He noted indeed
many examples in their entirety. How to get to quadratures was known to him; he merely indicated the process
by the sign of a sum, i. e., by the symbol of integration.
On the other hand, the part on the numerical solution of
adfected equations was new to him, and this he copied out
well-nigh word for word; this is the well-known Newtonian method of solution of equations by approximations.
Leibniz passes over as well known to him the remark, made
by Newton at the close of the quadratures, that the problems of rectification, determination of the content of solids,
determination of the centers of gravity, can be solved in
the same way, and also the general indication of the process
to be followed in such cases. Then follows the solution of
inverse problems, for instance, to find from the area the
base, that is the axis of the curve. This Leibniz copied
out word for word. In the same way Leibniz has extracted
the conclusion of Newton's paper, *"Demonstratio resolu-
tionis aequationum affectarum."* At the end of his manuscript Leibniz adds: "I extracted this from the letter of

Newton, August 20, 1672, addressed to Newton."[24] Probably this means that from the letters referring to Newton, Leibniz picked out the letter dated August 20, 1672, addressed to Newton.[25] So far as the script can be deciphered,[26] its contents were a graphic representation of Newton's method of solution of equations by approximations by means of Gunter's scale. Gunter's line had been noted by Leibniz on his first visit to London.

Of quite special interest to Leibniz were the letters of mathematicians which Collins had collected; on a second folio he made excerpts from letters from James Gregory. In two letters from Gregory (1670) was Isaac Barrow extolled as the greatest, not only among living writers, but also among all those that had written before him (Barrow). Further Leibniz found among these letters the letter mentioned above of Newton to Collins of December 10, 1672;[27] he extracted what Newton had mentioned with regard to his method of finding the expression for the tangent to a curve. Leibniz added at the end of this extract, "This method differs from that of Hudde as well as from that of Sluse, in that irrationals need not be eliminated."[28]

[24] The Latin, "*Excerpsi ex Epist. Neutoni 20 Aug. 1672 ad Neuton*," as given by Gerhardt, seems somewhat unintelligible; especially the word *Neuton*. What Collins had (or what Oldenburg, as suggested by Sloman, had) was a copy of a manuscript that Newton had sent to Barrow. Gerhardt says, "so far as the script can be deciphered"; perhaps the word *Neuton* is an error of transcription, or maybe an error on the part of Leibniz, due to the juxtaposition of the *Neutoni* which comes just before. In any case, Note 25 applies.

[25] I do not think Gerhardt's translation of the word *excerpsi* is correct.

[26] Gerhardt does not state whether the extract is badly written (this would show that it had been done in a very great hurry, for Sloman says that Leibniz, in his matter for publication, wrote a beautiful hand), or whether spoilt by age; in the latter case, as old-time inks contained salts of iron, the manuscript might be restored by photography, by means of a special plate, that I understand is sometimes used for detecting forgeries in deeds and notes.

[27] The letter was sent to Barrow to be sent on to Collins, probably with the object of being communicated through the latter to others; Collins seems to have been the regular channel of communication at this period, in a similar way to Mersenne.

[28] So we find in a manuscript, dated July 11, 1677, first of all an allusion to Sluse's method of tangents, "in which the equation is purged of irrational or fractional quantities"; then the remark, "I have no doubt that the gentlemen

From these extracts it follows that the contents of Newton's letter were unknown to him at that time (Oct., 1676).[29]

Regarding the verbal communications that Leibniz had from Collins during the second stay in London, Collins wrote to Newton from London on March 5, 1677 (1676 O. S.), that the representation of the roots of an equation by a series was discussed between them.

It is clear that Leibniz during his second stay in London had made himself more familiar with the results obtained by English mathematicians than he was before. The question now arises: What specially occupied his attention? What had particular influence upon his studies? It is seen that what Leibniz found in Collins's collection relating to algebraical analysis was new to him and excited his interest; also the verbal exchange of ideas between himself and Collins was upon the same subjects.

On the other hand, as regards the infinitesimal calculus, Leibniz obtained nothing during his second visit to London; he had made a progress, by the introduction of his algorithm into the higher analysis, beyond anything that came to his knowledge in London.[30] Also these algebraical results, at least for the next period, left behind no lasting impression; for among Leibniz's papers is to be found an extensive treatise, written on board the ship that carried him from London to Holland, wherein he considered the

---

I have just mentioned know the remedy that is necessary to apply"; then follows the rule for a quotient, and the remark that this will be sufficient for fractions; lastly the rule for powers, with the remark that this will be sufficient for irrationals. Later, he says, "This method has more advantage over all others that have been published than that of Slusius over all the rest, because it is one thing to give a simple abridgment of the calculation, and quite another thing to get rid of reductions and depressions."

Thus, after the sight of Newton's paper, his whole business has been to improve the method of Sluse.

[29] I read it quite otherwise; he has had information of some kind, whether from Oldenburg direct or from Tschirnhaus, while in Paris, and visits London with the express intent of seeing the original papers.

[30] See the third critical note, p. 181.

fundamental principles of motion, in the form of a dialogue.[31]

It was in the letter to Oldenburg written from Amsterdam on November 18/28, 1676,[32] which Collins spoke of in the letter to Newton mentioned above, that Leibniz first refers to the subject of the problem of tangents, and remarked that the method of Slusius was not yet very perfect.[33]

[31] Could this possibly have had its rise in an effort on the part of Leibniz to understand fluxions, or rather the idea of fluxions as he had found it in Newton's paper?

[32] In 1582, Gregory XIII had directed 10 days to be suppressed from the calendar, then in accordance with the Julian system of intercalation, in order to allow the error which had crept into the time of the vernal equinox, by which Easter-day was settled, to be put right. The Gregorian calendar was introduced into all Catholic countries the same year, in Scotland in 1600, in the protestant states of Germany in 1700, but not in England until 1752. At the same time the commencement of the legal year in England was altered from May 25 to January 1; thus we frequently find two years given for dates between January 1 and May 25; while there are two days of the month given for all months of the year. For instance, February 1673 in the new Gregorian calendar would be only February 1672 in the Julian, distinguished by the letters O. S. (Old Style); and this date was written February $167^2/_3$. Similarly the date November $^{18}/_{28}$, 1676, was the 28th of November in the New Style, and the 18th in the Old Style, the number of the year being the same, since the day did not lie between the 1st of January and the 25th of May.

[33] "*Methodus Tangentium a Slusio publicata nondum rei fastigium tenet.*" These are Leibniz's words; Gerhardt omits to translate the word *publicata*, which probably refers to the publication in the *Phil. Trans.* of 1672, by Slusius, of the rules of his method, illustrated by examples. Sluse had probably improved upon this before 1676, but there is no evidence on this point. It would seem as if the subsequent work by Leibniz, culminating in the manuscript of July 11, 1677, was largely an attempt to perfect the rule of Sluse *as a rule*, and that Leibniz, if ever, did not appreciate the idea fundamental in the calculus, namely that of rates, until very much later.

CRITICAL NOTES ON GERHARDT'S ESSAY.

NOTE I. *The origin of Leibniz's "transmutation of figures."*

(Referred to in footnotes 12, 13, 15.)

In the manuscript, which follows under heading II, Leibniz appears to attach very considerable importance to the method of transmutation of figures, and to claim that he had originated it. This claim is not incontestible; indeed I am almost inclined to think it is a deliberate plagiarism to start with; but Leibniz has perceived in it something which the original author did not. Can it by any chance be the case that, in conformity with several other instances of Leibniz's bad memory for details, he is confusing author and subject, when he speaks of "the great light that suddenly dawned on him, which the author had missed," the reference being to Pascal and the discovery of the differential triangle? Can it be that the true connection is that in considering the original work of the author of such transmutations of figures, he perceived the method for the arithmetical quadrature? For here he really has found a thing that the author missed though it was almost staring him in the face, his discovery being due to a habit that Leibniz had of writing down everything that he could get out of any particular figure or bit of work that he had in hand, whether it was relevant or irrelevant.

Wallis and Pascal had both hinted at the method, i. e., had used it in special cases, namely for proving the equivalence of the parabola and the spiral; and Leibniz was familiar with both these authors. Again, James Gregory had, in the words of Barrow (*Lect. Geom.,* Lect. XII, App. 3, foreword to Prob. IX), "set on foot a beautiful investigation about involute and evolute figures," i. e., polar and rectangular figures equal in area to one another. Of course, Leibniz may not have seen this work of Gregory until later; probably not, although in one of his manuscripts he gives a theorem of Gregory; this however does not count for much, for the very same theorem is given by Barrow (see my *Barrow,* p. 130) and we know that Leibniz had a Barrow in his possession. This book, judging by his words, "as in Barrow, *when his Lectures appeared,* in which I found the greater part of my theorems anticipated," Leibniz wishes to make his friends believe was the 1674 edition, and not the edition of 1670, which he bought on his first visit to London. Why did Leibniz wish to conceal this fact? I assert that the reason for doing so was the fear that seemed always to overshadow him, the fear of being accused of plagiarism, whether such was a true or a false charge. I am firmly convinced that Leibniz got his transmutation of figures from Barrow; to this conclusion I have only just come, it never having entered my head to look for it at the time that I wrote my articles for *The Monist* of October, 1916, April and July, 1917.

Before I bring forward my arguments, it is right to state as a preliminary that, just as in calculus nowadays we usually draw a curve with its convexity downward, and draw the tangent to meet the horizontal axis beneath the curve, so Barrow drew his curves with the concavity downward in many cases, mostly, I think, in order to fit the diagrams conveniently on the old-fashioned folding plates of diagrams, that in those days were added in batches at the end of a book (see a specimen I have given at the end of my *Barrow*); in other cases, he drew his figure on the left-hand side of the axis. Whichever figure he drew, he always did one thing, namely, he drew any supplementary figure he had need of *on the other side of his axis or base*. Leibniz almost invariably drew his curve on the right-hand side of a vertical axis, and supplementary figures on *the same side*. Hence, in the extract from Barrow given below, I am to be excused for failing to notice before what is more than a mere similarity.

In the following extract from Barrow (Lect. XI, Prop. 24), I have added Barrow's proof, which I thought unnecessary to give in my book; the figures given are Barrow's own on the left, which has been "up-ended" on the right; the latter is to be compared with the several figures by Leibniz.

Barrow's Lectiones Geometricae, Lect. XI, Prob. 24.

If DOK is any curve, D a given point on it, and DK any chord; also if DZI is a curve such that when any point M is taken in the curve DOK, DM is joined, DS is drawn perpendicular to

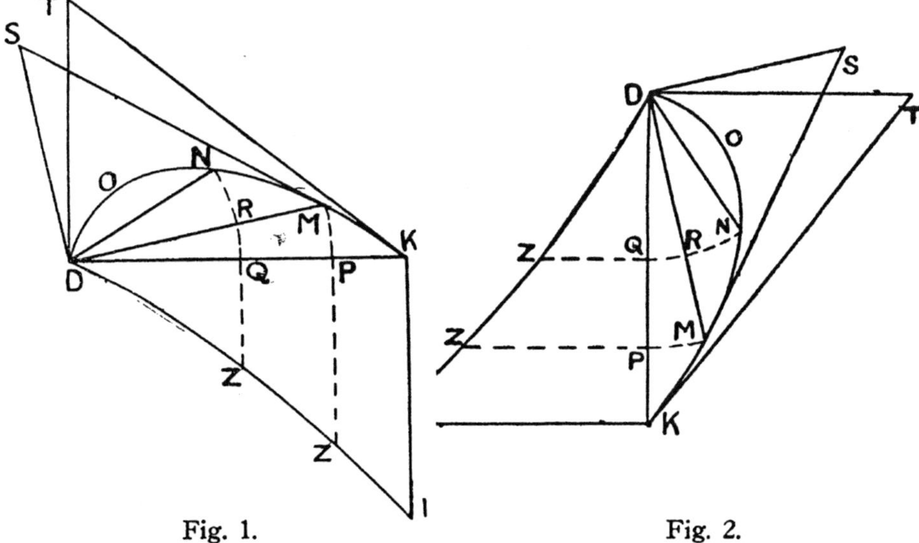

Fig. 1.Fig. 2.

DM, MS is the tangent to the curve, DP is taken along DK equal to DM, and PZ is drawn perpendicular to DK, so that PZ is equal to DS; in this case the space DZI is equal to twice the space DKOD.

For let KP be considered to be indefinitely small, and let DT be perpendicular to DK and KT the tangent to the curve DOK. Then, drawing the arc MP, we have *as before,*

KP:PM = KD:DT = KD:KI, and hence KP.KI = PM.KD. Take another small part PQ and, with center D, draw an arc QN through Q cutting the chord DM in R; then *as before,*

MR:RN = MD:DS, PQ:RN = MD:PZ, PQ.PZ = RN.MD; and so on one after the other. Therefore, it is evident that the sum of all the rectangles KP.KI, PQ.PZ, etc., is equal to the aggregate of all the spaces PM.KD, RN.MD, etc.; that is, the space DKI = 2 times the space DKOD.

The words I have italicized refer to Prop. 22, in which he uses a similar though rather more complicated figure to reduce a polar area to a rectangle *of which one side is a given straight line,* and explains that the reasoning depends on the fact that the line DK is divided into infinitely small parts. Compare the words I have italicized with the description of Leibniz's method: "the areas of these triangles *being proportional to lines.*

Further, Barrow proceeds in Prop. 25 to prove the equivalence of the spaces formed (i) by applying each MS to the base and (ii) by applying each chord to the arc, previously rectified. And he winds up with the words: "Should any one explore and investigate this mine, he will find very many things of this kind. Let him do so who must, or if it pleases him."

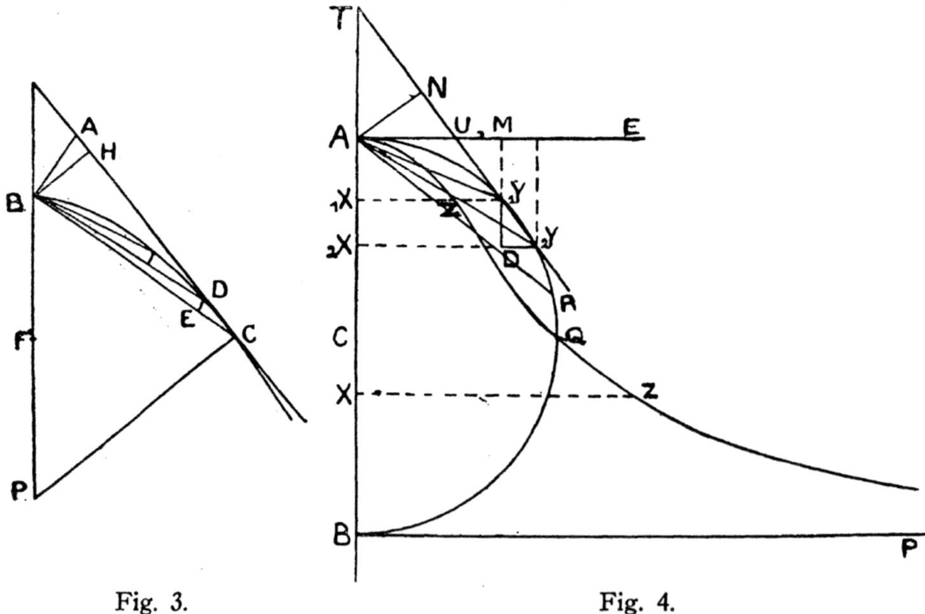

Fig. 3.　　　　　　　　Fig. 4.

This all suggests that Leibniz *did* explore this mine, that he did *not* invent the method of transmutation of figures for himself, that he *did* find very many things of this kind, and that Barrow had *missed the arithmetical quadrature* construction; this Leibniz obtained through his regular practice of working every mine right out, to keep up Barrow's simile. Further comment is needless, I think, after a comparison of Barrow's figure (the up-ended version) with the figures of Leibniz given above.

Fig. 3 occurs in a manuscript November 21, 1675, which according to Leibniz is at least a year after he had discovered the arithmetical quadrature; and yet it has a heading, "A new kind of Trigonometry of indivisibles, etc." In this figure it is to be noticed that he has the perpendicular to the chord BC, agreeing with Barrow's DS and DT, but has not the tangent at the vertex that was necessary for the demonstration of the arithmetical quadrature. In the working in connection, he considers the similarity of all the triangles possible, and notes as one point that *"the sum of all the triangles or the area of the figure is equal to the products of the AB's into the CE's,* which is Barrow's proof of Prop. 24 above.

Fig. 4 is the figure given in the *Historia* (see above, Chap. III. p. 42) in connection with the explanation of how he found the area of the circle. Notice the difference between this figure and the one given in the manuscript that follows under the heading II, also that the description there given of the way in which he was led to it is much more natural. This is probably the true version, for the use of the notation B, (B), ((B)), points out that it was written at a comparatively early period, before Leibniz had adopted the prefix notation, $_1$B, $_2$B, $_3$B. In the account in the *Historia,* to which Fig. 4 applies, Leibniz says, "he once happened to have occasion to break up an area into triangles formed by a number of straight lines meeting at a point, and he perceived that something new could be readily obtained from it." I suggest that the occasion was most probably while he was digging in Barrow's mine! This is the reason why he has in the *Historia* given the figure more according to his usual practice, and different from the figure in the earlier manuscript, which is too much like a copy of Barrow's (query, where did Barrow get it from?). With regard to the figure and proof in the manuscript which follows, we find that the reasoning there given is unsound, unless Gerhardt has given us a slightly erroneous diagram; for Leibniz apparently does not perceive that the ordinates BA, which are equal to the corresponding CE, *must pass through the respective points D,* before he can say that one figure is double the other. Hence I conclude that at the date of this manuscript, the demonstration was imperfect and that he had no proof until he dug in Barrow's mine; in support of which conclusion I will quote from the *Recensio,* mentioned in Note 22, p. 167: "This quadrature, composed in the common manner, he began to communicate at Paris in the year 1675. The next year he was polishing the demonstration of it, to send it to Mr. Oldenburg, in recompense for Mr. Newton's Method, as he wrote to him May 12, 1676; and accordingly in his letter of August 27, 1676, he sent it, composed and polished in the common manner." This polishing, I· take it, consisted in making the slight but important alterations in the demonstration and figure, from those given in the manuscript II that follows, to those given in the *Historia*.

What had he then got in July 1674, when he wrote to Oldenburg saying that he had got a wonderful Theorem, which gave the area of a circle, or any *sector of it exactly,* in a series of rational numbers? Or, when in the October following, October 26, 1674, he wrote to say that he had found the circumference of a circle in a series of very simple numbers; and also by the same "method" (a favorite expression of Leibniz) *any arc whose sine was given?* It was impossible that Leibniz could have had the two things that I have italicized; or at least, the latter was impossible to him, because the only way for him to obtain it *exactly,* i. e., to know the law of his series, was as yet unknown to him; unless we are to assume, contrary to his assertion, that the binomial theorem was known to him, which would involve his also having seen or been told about other parts of Newton's work. The only way open to Leibniz was to find the square root of $1-x^2$, and then its reciprocal by division; and this would not give him the law of the series, even if we assume that his knowledge of integration was sufficient to enable him to proceed any further. From his manuscripts it does not seem that even up to Nov. 1675 he had any further knowledge of integrations than that omn.$x = x^2/2$, and omn.$x^2 = x^3/3$; but as he says that he knows the latter from the quadrature of the parabola, there is some possibility that he might have been able to integrate every integral power of the variable from his reading of Wallis and Mercator.

However, there is the strongest probability that he had not got any proof for the two things italicized, and that the quadrature was in the same category. Where then had he obtained it? We find that in December, 1670, Gregory had found out for himself Newton's method of series; and two months later, February 15, 1671, sent several theorems to Collins, one of which was that now known as "Gregory's series." "And Mr. Collins was very free in communicating what he had received both from Mr. Newton and Mr. Gregory, as appears by his letters printed in the Commercium" (from the *Recensio*). One can imagine that Oldenburg would be one of the first to receive the information, and that for a certainty it would be passed on to Leibniz. I think then that Leibniz perceived that by putting $x = 1$ in Gregory's series, and making the radius of the circle equal to unity, he could get an arithmetical quadrature; from that time onward he looked for a proof by pure geometry, and found it after reading Barrow's proposition referred to above; if we assume the possibility of integration of integral powers, it was an easy step to find that the series he had to integrate was $y^2/(1+y^2)$, and all he had to look for on his figure was a line of this length. This very well accords with the description of the way in which he found his demonstration, as given in the manuscript which follows under the heading II.

Lastly, in connection with the suggestion that I have made above, namely, that Leibniz had another method for his arithmetical quadrature than those he has given, there is one method that is bound up with the change that he made from the Pascalian characteristic triangle which he used at first, to the Barrovian differential triangle (see a note on this point, Chapter I, p. 15). In Example 5 of the method of the differential triangle (see my *Barrow*, p. 123), Barrow has found the subtangent for the curve $y=\tan x$, from a consideration of the figures below, and finds that

$$t=\frac{rr}{rr+mm}m.=\frac{CB^2}{CG^2}.BG=\frac{CK^2}{CE^2}.BG\ .$$

where r is the radius of the circle, m is the ordinate MP, which is equal to BG, and t is the subtangent TP.

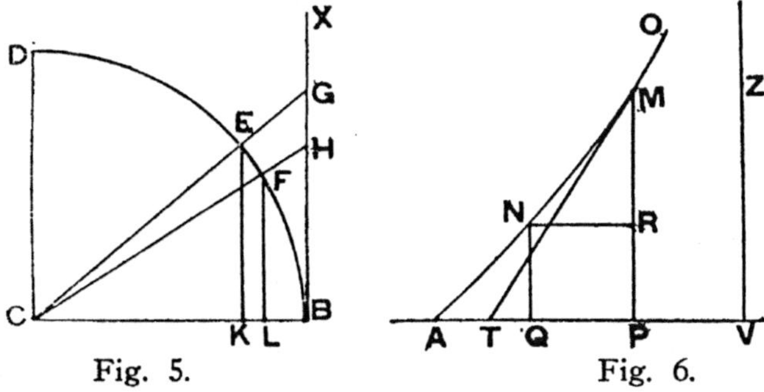

Fig. 5.   Fig. 6.

Now if we put the radius equal to unity, and for the ratio t/m substitute what was known by Leibniz to be equal to it, namely, QP/RM or EF/GH (by construction), we have the sum of all the EF's is equal to the sum of ordinates equal to CK^2 (radius$=1$) applied to G at right angles to BG. Analytically, calling BG z, we have

$$\text{arc BE}=\text{sum.omn.}\frac{1}{1+z^2}\ \text{applied to the line}\ z\ ;$$

hence by division

$$\text{arc BE}=\text{sum.omn.}(1-z^2+z^4-z^6+\text{etc.})$$
$$=\quad z-z^3/3+\text{etc.}$$

I can hardly see how Leibniz could have missed this with his analytical mind, even although Barrow has missed it; but there is a strong probability that at the time of writing, Barrow had not seen the quadrature of the hyperbola by Mercator, and, if he had, such algebraical work would not have appealed to him at all.

As far as I can make out, there is only one other alternative, which involves a direct contradiction of Leibniz's own statement; that is that his proof was not by the transmutation of figures in the first instance. Color is lent to this view by a letter of Leibniz and other papers, quoted by Sloman (pp. 131ff, in the English edition of the work referred to in footnote 23) ; also even by a passage in the *Historia* (see *Monist*, Oct., 1916, p. 599), where, while giving the story of the discovery of the arithmetical tetragonism, Leibniz distinctly hints at an algebraical method ; for he says immediately afterwards, "The author obtained *the same result by the method of transmutations,* of which he sent an account to England." This reads as if he had another method in addition to the method by transmutations.

Let us consider this algebraical method. To square the circle, Leibniz has to integrate $\sqrt{(1-x^2)} = y$, say ; let $y = 1 - xz$, then $y = (1-z^2)/(1+z^2)$, which is rational ; moreover, he would also have been able to have substantiated his statement that at this time he also had a proof of the series for the arc whose sine was given, for which he would only have had to integrate $1/\sqrt{(1-x^2)}$. But one cannot conceive that Leibniz had any means of expressing the element of z in terms of the element of x. Geometrically, he was incapable of it, without using Barrow's infinitesimal method ; and of this we find the first instance in a manuscript dated November 1, 1675. Algebraically, he could not, for at this same date he could not differentiate a product. How then are we to account for the fact that he says he has a method for demonstrating both series for the arc, given the sine or the tangent? I think I can answer this. Many times we find assertions made, not only by Leibniz in those times, but by others in other times, of the possession of discoveries, when all that the assertor has is the idea of how they may be obtained. Thus, in the passage quoted, the concluding statement is, "and thus again all that remains to be done is the summation of rationals." So that if we accept this alternative we are bound to come to the conclusion that Leibniz did not yet recognize, what he ought to have done from the work of Pascal, that an area was not a mere summation of lines, but of rectangles formed by these lines ordinated at certain definite points along a straight line. That is to say, he did not recognize the fundamental principle of integration, namely, the importance of the factor dx or dz. When he had to write out his proof he found that the summation of $(1-z^2)/(1+z^2)$ or its reciprocal was beyond him ; or rather that the series he found by Mercator's method was not correct ; he had to resort to the geometrical proof, of which he got the idea by digging in Barrow's mine, as above ; he found that this would not work for the other series ; and consequently he dropped all claim to the second series. In his letters of 1676, therefore, we find him offering to send Newton the proof of his quadrature in return for the method of proof of the series for the arc when the sine is given.

Thus I come to the conclusion that Leibniz obtained these series in some way by correspondence, thought he had got a proof of his own, (which turned out to be incorrect), and much later did obtain a proof of his arithmetical quadrature by the transmutation of figures, *after obtaining the idea from Barrow.* As the special case, when $x=$ the radius, had not been specifically mentioned by Gregory, Leibniz considered that he had a right to claim it, more particularly as he thought he had devised a proof for it, if it was necessary to produce one; for of course, Gregory had given no proof according to the usual custom of the time. Then, when he did find a proof, after having found that his original idea was hopeless, one can hardly blame him for sticking to his claim.

NOTE 2. *On the introduction of the Leibnizian algorithm.*
(Referred to in footnote 21.)

The two passages in which the signs for integration and differentiation are respectively introduced occur in the manuscript of October 26, 1675.

i. "It will be useful to write $\int$ for omn., so that $\int l=$ omn. l, or the sum of the l's."

ii. Not for some time is the sign for differentiation introduced, and then in these words: "I propose to return to former considerations. Given l and its relation to x, to find $\int l$. Now this comes from the contrary calculus, that is to say if $\int l=ya$. Let us assume that $l=ya/d$, or as $\int$ increases, so d will diminish the dimensions. But $\int$ means a sum, and d a difference. From the given y, we can always find ya/d or l, or the difference of the y's. Hence one equation may be changed into the other,......"

Now of these the introduction of the symbol for integration can no more be called an invention than the use of Σ to stand for "the sum of all such terms as." It was simply, as Leibniz himself says, a convenient and useful abbreviation for sum.omn. or omn. It is nothing more or less than the long s then in general use; indeed it was so thought of by contemporary mathematicians, Newton for one at any rate, for we find in the *Recensio* the passage, "Mr. Leibniz has used the symbols sx, sy, sz for the sums of ordinates ever since the year 1686." This may have been an instance of prejudice, or perhaps the printers of the *Phil. Trans.* may not have had an integral sign in their fonts of type; but it shows up the fact that the English accepted it as the initial letter of the word "summa."

Now let us consider the introduction of the letter d. Gerhardt says that it resulted as antithesis to the sign $\int$. How he can possibly derive this from the context I cannot surmise. I am well aware that in another passage he was unable to assign a meaning to the introduction of a letter, which was, to me, clearly used for the simple purpose of keeping the dimensions correct. We have this use again in the present passage. Leibniz knows that the sum of the lengths, $\int l$, is an area; hence taking y to represent a length, given in terms of x, he introduces the *length* denoted by a to give with y the area of a rectangle. Therefore he argues that l must be an area divided by a length, and he writes $l=ya/d$, where d is *another length, intro-*

duced to keep the dimensions correct. This is clear from the sentence that follows next: "so will *d* diminish the dimensions."

So far the sequence of ideas is easy to follow, and there is not the slightest trace of any concept of differentiation, nor, if the *l*'s are ordinated to any axis, any trace of a connection between *d* and an element of that axis. The difficulty begins with the next sentence: "But *∫* means a sum, and *d* a difference." The first idea that strikes one is that this was added later, after that he had found out the connection between the inverse-tangent problem and quadratures. Gerhardt gives no suggestion on the point, so until the paper can be reexamined for small details like differences in the ink or character of the writing this idea will be disregarded. The next is that about this time he was reading Barrow, and then one is at once reminded of Lect. X, Prop. 11; this is the proposition in which Barrow proves that differentiation is the inverse of integration. If we consider this in the manner of Leibniz, we get the equivalent that is set down on the right-hand side below:

Let ZGE be any curve of which the axis is VD; and let ordinates applied to this axis, VZ, PG, DE, continually increase from the initial ordinate VZ; also let VIF be a line such that if any straight line EDF is drawn perpendicular to VD, cutting the curves in the points E, F, and VD in D, the rectangle contained by DF and a given length R is equal to the intercepted space VDEZ; also let DE : DF = R : DT, and join DT. Then TF will touch the curve VIF.

Cor. It should be observed that DE . DT = R . DF = area VDEZ.

Let AC be a curve, whose axis is AB, and let the ordinate AB be *l*;

let AD be another curve, having the same axis, and let its ordinate DB be called *y*;

let this curve AD be such that the area ABC, i. e., all the *l*'s or *∫ l*, is equal to the product of BD and a fixed line, i. e., equal to *ay*;

then, taking B(B) equal to unity, we have $l = aw$, where $w : B(B) = DB : BT$, or $w = y/d$, i. e.,
$$l = ay/d.$$

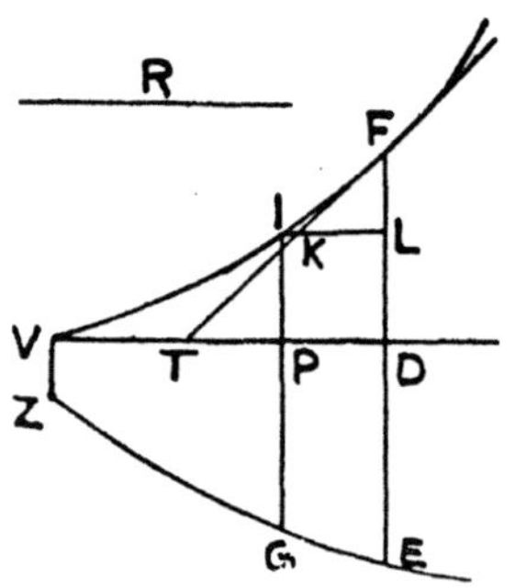

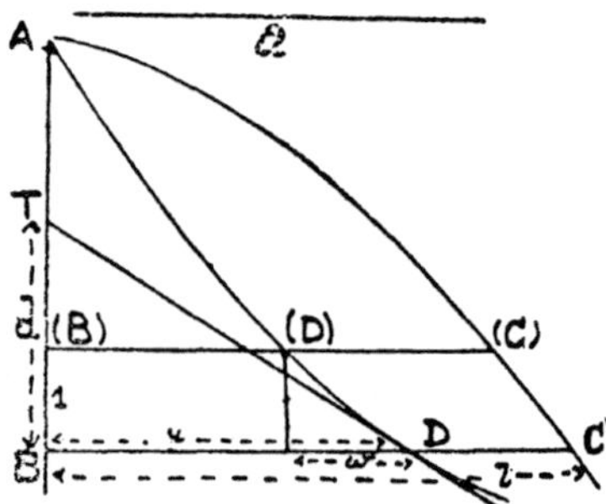

We thus see that the d that results as the "antithesis to the integral sign" (*als Gegensatz....sich ergab*), is not a difference at all, *but the subtangent*; it is y/d or w (on account of B(B) being taken as unity) that is the difference between the ordinates y. But there is not the slightest trace of the idea of differentiation; this is made more manifest by the work which follows, which is based on his idea of obtaining independent equations, and eliminating all variables but one and thus reducing the problem to a quadrature. And yet he seems to perceive from the equation that gives the difference of the y's as a *quotient,* that in some unintelligible way a division means a difference. Later therefore we find him trying to find an interpretation of d as an *operator,* whether he writes it in front of his y, or as a denominator; namely, when he considers what value he is to assign to $d(xy)$. I venture to assert, unless we assume that Leibniz is considering this proposition of Barrow's, that there is no possible connection to be made out between the several sentences of this passage. Also that in no sense can this introduction of the letter d be looked on as anything connected with an algorithm with any idea in it of differentiation.

I am well aware that in the above I have adduced no positive proof that my idea is correct; I have not had the advantage of Gerhardt in seeing these manuscripts. But I have honestly tried to find other ways of explaining the circumstances that lead from y/d as a *quotient* to dy as a *difference,* and I can find none other that is feasible than that given above, namely, that, perhaps by accident, Leibniz uses d for the subtangent (instead of the usual t), and perceives from such a figure as the above (which of course I do not intend to say he has given) that y/d (where d is the subtangent) works out the same as dy (when dx is taken to be unity); in other words the subtangent d is equal to $y/(dy/dx)$.

NOTE 3. *On the progress made by Leibniz before November, 1676.*

(Referred to in footnote 30.)

The remark made by Gerhardt that Leibniz "had made a progress, *by the introduction of his algorithm into the higher analysis,* beyond anything that came to his knowledge in London," is, to say the least of it, a matter of opinion. From a study of the six manuscripts, that Gerhardt has given us, that bear dates between that of the introduction of the integral and. differential symbols (Oct. 26, 1675) and that of his return to Germany, via Amsterdam (after Nov., 1676), I fail to see that there is very much occasion for the main part of the above statement, namely, that the progress made by Leibniz was at all greater than anything that came to his knowledge in London; as for this progress, if for a moment we **assume its superiority, being due** to the reason set in italics, I fail to see that Gerhardt has any grounds whatever for such a statement.

The six manuscripts in question have been given, translated into English and annotated, in Chaps. III and IV, pp. 84-121; for convenience I here add a précis of them.

i. Nov. 1, 1675. A continuation of the work on moments about axes; the new symbols do not occur, *omn.* being still used. He has now read Wallis, Gregory and Barrow, in addition to Cavalieri and St. Vincent; he speaks of his theorem of breaking up a figure into triangles as bringing out something new; the whole tone of this manuscript is in the main Pascalian.

ii. Nov. 11, 1675. He successfully obtains a solution of the problem of finding a curve such that the rectangle contained by the subnormal and ordinate is constant. This he considers to be "one of the most difficult things in the whole of geometry." He uses the integral sign, and the denominator d; but neither integration nor differentiation, the fact that $y^2/2d = y$, being taken from the "*quadrature of the triangle.*" In verifying his result he quotes Slusius's Rule of Tangents. Further on, he has the note that x/d and dx are the same thing, though there is nothing to show why he comes to this conclusion; see the last critical note. He also comes to the conclusion that $d(xy)$ is not the same as $dx.dy$; but in the last bit of work in this manuscript he uses special letters for the infinitesimals, showing that he has been trying to find the effect of d as an *operator,* or perhaps trying to find the reason of the equality x/d and dx. He has failed to solve a problem, which results in the differential equation, as we should now write it, $x + y.dy/dx = a^2/y$, or as Leibniz has it $x + w = a^2/y$; although he gives an incorrect solution, which he asserts to be true. This time he does not attempt to verify his solution, the reason being obviously that he is unable to do so, because one side of his equation is a product. As a matter of fact, I have it on the authority of Professor Forsyth that there is no solution of this equation in elementary functions; or at least he says that he has been unable to find one, which I take it comes to the same thing. The one advance that can be found here is the appreciation that squares and products of infinitesimals can be neglected, as he has doubtless found in reading Barrow. It is worth noting that he now uses the differential triangle in Barrow's form instead of the form he says he got from Pascal.

iii. Nov. 21, 1675. In this manuscript he sets himself another problem, which he fails to solve; the curve required is logarithmic, and even this fact he fails to bring out. In generalizations that arise from the consideration of his problem he obtains $dx\,y = xy - x\,dy$, in a more or less analytical manner; but immediately afterward states that nothing new can be obtained from it; he has already obtained this formula by his consideration of moments, geometrically; and he does not appreciate the advance there is in obtaining it algebraically. The manuscript concludes with a consideration of the figure by means of which it is generally supposed that he effected his arithmetical quadrature. This is very remarkable on account of the heading, which reads, "A *new* kind of Trigonometry of indivisibles, by the help of ordinates that are not parallel but converge." What I refer to is the use of the word new, which I have here italicized. It is to be observed that the diagram and the results are almost identical with those of *Barrow,* Lect. XI, Prop. 22-24 (see the first critical note). He concludes by a reference to the trochoids. which shows that he is still under the influence of Pascal, if indeed he is not still studying his works.

iv. Nov. 22, 1675. He returns to the subjects of the previous day. But there is here no mention of the signs of integration or differentiation.

v. June 28, 1676. Here we have a certain advance, for there occurs the statement: "The true general method of tangents is by means of differences." While he uses dy and dz for the elements of y and z, he uses β for the element of x; the rest of the work is merely Barrovian in principle. This mere substitution of dy and dz for the special letters used by Barrow for the same things can hardly be called progress. What progress there might be is barred by the use of equations with three or more variables in them.

vi. July, 1676. The remark on the last manuscript is corroborated by the contents of this manuscript. Leibniz asserts that he has solved two problems, of which Descartes had alone solved one, and owned that he could not solve the other. The truth is that he has not solved the former, which was fairly easy, only given an alternative construction which is, if anything, more difficult to carry out than a construction from the original data for the curve. The latter he gets out in a hazy fashion (". . . . which belongs to a logarithmic curve"). This conclusion he comes to after several erroneous steps of reasoning; whereas the solution stared him in the face about a quarter of the way through the work, where he has the equation $c\,dy = y\,dx$, if he could have integrated dy/y with certainty. The failure I think arises from the study of Pascal, who lays it down that only one of the variables can increase arithmetically, and Mercator's work has been with y increasing arithmetically, and Leibniz has already considered that the x is increasing arithmetically. (See Note 55 on this manuscript above, Chapter V, p. 121.)

Throughout the whole of these manuscripts, he makes no progress, because he is hampered by the idea of keeping one of his variables increasing uniformly; he seldom uses his algorithm for differentiation; and when he does do so, it is merely a substitution of dx, etc. for the special letters used by Barrow. In fact these manuscripts appear to me to be the records of his work on the textbooks of his study, Pascal, Wallis, Gregory, and Barrow; and we see him trying to fit the matter and methods found in them into his own ideas and notations. It is not until November, 1676, when he has arrived on the Continent, after having seen Newton's paper, that we have any Differential Calculus; even then some of the standard forms that he gives are not quite correct; on the other hand, he gives the method of substitution to differentiate an irrational, though he uses the Barrovian method to differentiate the general equation of the second degree, merely using dy and dx instead of Barrow's special letters. It is not until July, 1677, that he is able to give anything like an intelligible account of the differentiation of products, powers, quotients and roots. Lastly I doubt if Leibniz ever did really appreciate the Newtonian idea that dy/dx was a *rate*, or else the example he gives of the use of the second and third differentials in his answer to Nieuwentijt would not have contained so many ridiculous errors.

TRANSLATIONS OF THE MANUSCRIPTS

Alluded to by Dr. Gerhardt.

I.

Scientific memoranda of the visit to England at the beginning of the year 1673.

When at the beginning of the year 1673, I accompanied his Excellency the Ambassador of Mainz, Baron Schönborn, a nephew (on his father's side) of the Elector, from Paris to London, although I stayed in England scarcely a month, among various distractions, I still gave attention to increasing my knowledge of philosophy; for at that time the English held a high reputation in this subject.

To set out a long minute record of daily happenings is useless on account of its inequality; for the fortune of all the days was not the same; indeed the points worth remarking heaped themselves up one day, and the next gaped with emptiness. For this reason perhaps it will be more satisfactory to go by heading of subjects, one remark recalling another as it were.

The principal heads for the subjects noted may be taken as Arithmetic, Geometry, Music, Optics, Astronomy, Mechanics, Botany, Anatomy, Chemistry, Medicine, and Miscellaneous.

ARITHMETIC. The line of proportions or Gunter's lines or the double scale. Logarithmotechnia or compendium for calculating logarithms. To recognize square numbers from non-squares by their end figures. Morland's machine.

ALGEBRA. Substance of English algebraical work of 27 years. Algebra of Pell. At first few rules, but numerous selected examples. Renaldinus not thought much of in England.

GEOMETRY. Tangents to all curves. Development of geometrical figures by the motion of a point in a moving line.

MUSIC. Its universal character. System of Birthincha(?). Vossius will publish Music.

OPTICS. They told me of a certain phenomenon that Barrow confessed that he was unable to solve. The difficulty of Newton hitherto unsolved, Father Pardies giving it up. Hook adheres to a catadioptric instrument of 9 feet, because for another of 50 feet movement inconveniences them. The secret of the largest aperture which can be given to microscopes is primarily as great as the distance of the object.

| ASTRONOMY. Arrangement of Hook for observing whether the earth at any time sensibly approaches or recedes from the fixed stars, from which it can be judged that it is not in the center of the universe; he erected it in a fine tube set perpendicularly, and observed the stars that are vertically overhead. He, lying flat on his back, observed their dimensions most exactly. | CHEMISTRY. |

MECHANICS.

PNEUMATICS.

METEOROLOGY.

HYDROSTATICS.

NAVIGATION.

MAGNETISM.

PHYSICS.

BOTANY.

ANATOMY.

MEDICINE.

MISCELLANEOUS.

II.

[This manuscript is very lengthy, the translation running to about 6000 words, of which the first 5000 are written as a concise history of all the great geometers and their works, that are antecedent to Leibniz himself. This part is quite unimportant for the purpose of estimating the part that was played by Leibniz, and it passes my comprehension why Gerhardt should give it at length, while he has condensed the other two, which are really important. Hence, in what follows, I have given a precis of the first 5000 words, with here and there quotations, in which Leibniz has something to say that is either critical of the work of others, or a claim to superior knowledge or better method of his own. The last part, which purports to be the history of his arithmetical quadrature, together with his claim to the surpassing value of his achievement, I have given in full.]

(Precis). Geometry is a modern thing, probably due to the Greeks. The great name among the Ancients is that of Archimedes, who first used indivisibles; this use was more profound than that of Cavalieri, but the method became lost. The name of Apollonius must not be altogether omitted.

The learning of the Greeks passed on to the Arabs, who conquered them; among these we have Alhazen, and a certain Mahomet, who gave the formula for the general quadratic.

This brings us to the cubic and biquadratic equations, which were solved in the sixteenth century. The cubic is due to one Scipio Ferreus of Bologna; one of his pupils set the solution as a challenge after Scipio's death; Tartalea took up the challenge, found a solution and told his friend Cardan; the latter extended it and published it without the consent of Tartalea. Vieta, Descartes, and Ferrarius gave the solution of the biquadratic. But even Descartes and Vieta failed at equations of higher degrees. With regard to the work of Descartes, Leibniz remarks that "its origin [that is, of the method of solution] was a widely different and more fertile spring; and if Descartes had only recognized this, he would have rendered the discovery of Scipio more general and carried it to further heights. *But what has befallen me in this connection I will say in another place.*" Leibniz further remarks that the method of Descartes fails to give the roots of equations of higher degree, although the quality of the roots may be learned through it. "I will show in another place that the *reason for this is clearly known to me from the most fundamental principles of the art, and that I have established an extremely easy method, and one that is adapted too for enlarging science, by the many things that follow from it.*"

In the seventeenth century, Leibniz goes on to say, after Archimedes and Galileo's several times and influence are gone by, there is no writer from whom more is to be learned than from Descartes; and yet he is *"unable to pass over certain boastful remarks that he makes, by which the less experienced among us may be led into error."* Descartes had said that by his method every geometrical problem could be reduced to the finding of the roots of equations. Leibniz remarks that this shows Descartes's ignorance of the matter. "For when the magnitude of curved lines or the space enclosed by such is required (which happen more frequently than perhaps Descartes thought, since he had not applied himself sufficiently to the 'mechanics' of Galileo), neither equations nor Cartesian curves can help us, and *there is need of equations of a totally new kind, of constructions and new curves, and finally of a new calculus, given so far by nobody, of which, if nothing else, I can now give certain examples at least, which are remarkable enough."* .... *I have mentioned these things so that men may understand that there are certain methods in Geometry, for which they may look in vain in the works of Descartes."*

Returning to geometry purely, Leibniz next mentions the work of Galileo, Cavalieri (whose method he considers is rough and limited in extent), Torricelli, Roberval, Pascal, Wallis, Huygens, and Slusius, as contributors to the new geometry. He considers that a new epoch opens with the work of Neil and van Huraet (on rectification of curves), James Gregory, and Brouncker. "Finally Mercator gave a general formula for the area under a hyperbola." He claims Mercator as "an eminent German geometer"; but rather decries his discovery as being an easy one, on account of the ordinates working out as rational in terms of the abscissa. "But it was not so easy to give the magnitude of the circle, and its parts, expressed as an infinite series of rational numbers;.... for the circle, however you treat it, has ordinates that are irrational. However I, as soon as I had found a certain very general theorem, by means of which any figure whatever could be converted into another that is quite different from it, but yet of equivalent area, set to work to try whether the circle could not be converted in some way into a rational figure; and the thing came out beautifully;.... it will be worth while here to give a short account of the matter."

(In full). Nearly everybody who has up to now treated of the geometry of indivisibles has been accustomed to break up their figures into rectangles or parallelograms only by means of ordinates parallel to one another. But the reasoning of Desargues and Pascal always pleased me very much; these in Conics, as we can call them in general, include under the name of ordinates not only parallels, but also straight lines meeting in or converging to a point, especially when parallels are included under the name of converging, by saying that the point of convergence goes off to an infinite distance. Thus while others only consider parallel ordinates, and have broken up their figures into parallelograms AB(B)(A), (A)(B)((B)) ((A)), in the way that Cavalieri does, I employ converging lines and resolve the given figure into triangles CD(D), C(D)((D)), and at once draw another figure of which the ordinates AB, (A)(B), etc., are proportional to these triangles.

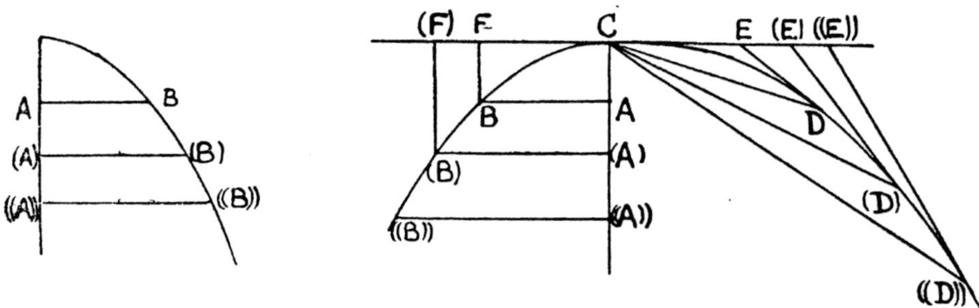

Now this is the case if the AB's are equal to the CE's where it is supposed that the straight lines DE are tangents to the given curve; for in that case, as I will show below, it will come out that the space B(B)(A)A will be double of the segment C(D)DC, and for any figure such as C(D)DC another that is equivalent to it can be drawn. Now, supposing that the curve D(D)((D)) is circular and that CA is a part of the diameter, then, calling CA or FB x, and CF or AB y, and the radius of the circle unity, calculation will show that the value of x is $2y^2/(1+y^2)$. Thus the ordinate FB or x can be expressed rationally in terms of the given abscissa CF or y. Such figures as these, in which the ordinates can be expressed rationally in terms of the abscissae, I call rational. Thus we have drawn a rational figure equivalent to the circle, and this will be soon seen to be sufficient to give the arithmetical quadrature of the latter. For, from the sum of a geometric series of an infinite number of decreasing terms that is well known to all geometers, it follows that $y^2 - y^4 + y^6 - y^8 + y^{10} - y^{12} +$ etc. to infinity is the same as $y^2/(1+y^2)$, i. e., the same as $\frac{1}{2}x$, if only we understand that y is a quantity that is less than the radius, or unity. Now, since we have to collect together the infinite number of $\frac{1}{2}x$'s into one sum, in order to obtain the quadrature of half the figure C(F)(B)BC and what it comes to, namely, that of the circle; so also have we to collect together the infinite number of series $y^2 - y^4 + y^6 - y^8 + y^{10} - y^{12} +$ etc., into one sum, and this by the method of indivisibles and infinites can be done without difficulty. For, suppose that the last y, which in general is taken as C(F), to be b, then the sum of every y^2 will be $b^3/3$, and of every y^4 will be $b^5/5$, and of every y^6 will be $b^7/7$, and so on; hence, the sum of the infinite number of $\frac{1}{2}x$'s, or of the series $y^2 - y^4 + y^6 - y^8 + y^{10} - y^{12} +$ etc., i. e., the area of half the space C(F)(B)BC, will be $b^3/3 - b^5/5 + b^7/7 - b^9/9$ etc. From which, by the help of ordinary geometry, it can be easily deduced that the square on the diameter is to the area of the circle as 1 is to $1/1 - 1/3 + 1/5 - 1/7 +$ etc.; also speaking in general, supposing b to be the tangent, then the arc is $b/1 - b^3/3 + b^5/5 - b^7/7 + b^9/9 - b^{11}/11 +$ etc. Hence it now follows that any one without the help of tables and continual bisections of angles and extractions of roots can approximate to the magnitude of the arc to any degree of accuracy desired, so long as the tangent b is a little less than the radius; so that if we take the tangent to be a little less than the tenth part of the radius, the arc may be obtained with sufficient accuracy. Let us take the tangent to be a tenth part of the radius, then if we want the arc, it will be

$$\frac{1}{10} - \frac{1}{3000} + \frac{1}{500000} - \frac{1}{70000000} + \frac{1}{9000000000} = \text{etc.} ;$$

and reducing all to a common denominator, and adding the numbers into one sum (for it is not worth while going any further), then the arc will be a little greater than 518027821302775/5197500000000000, and the defect of this value from the true value will be less than the 1/1000000000000 part of the radius. For if we do not subtract the last term, 1/1100000000000, the value would be too great, and if we do subtract it, the value is less than the true value, therefore the error is less than 1/1100000000000, and thus is less than 1/1000000000000.

It is seen how exactly it comes out with such easy calculation involving only additions, subtractions and multiplications, to an extent that is not obtainable with tables. Also if the ratio of the tangent to the radius is anything else, the arc can similarly be found, and this is especially easy when it can be expressed in decimal parts. Again, since now the ratio of the circumference to the radius is given in numbers of any required degree of accuracy, by this also the ratio of a given arc to the circumference is given, and thus also the quantity of angle for a given tangent will appear with any required degree of accuracy. In this way tables may be corrected, supplemented, or, if need be, enlarged, with no great trouble. Any one who will just remember this fairly easy rule will be able without tables to attain to any required degree of accuracy with very little labor. How great an acquisition this is to geometry, I leave it to those who understand to estimate.

CRITICAL NOTE.

It is difficult to see the object that Leibniz had in writing this long historical prelude to an imperfect proof of his arithmetical quadrature, unless it can be ascribed to a motive of self-praise. This suggestion would seem to be corroborated by the claims that Leibniz makes in the parts where I have quoted his own words in italics in the precis, and by the concluding sentence of the translation given in full. Even if this is so, there may be some plea of justification put forward; for Leibniz appears to have been a man impelled by many contradictory motives, but these I think can all be traced back to one origin. The time in which he lived was a time of great discoveries in geometry; Leibniz knew in his soul that he had it in him to be one of the great men in this branch of learning, but as truly recognized his great disability due to his lateness in starting, and felt that his only chance was to belong to the very exclusive set who corresponded with one another; he saw that the only way of entering this set was to do something brilliant. This may be taken as some excuse for any self-praise that we find, and to a less extent for his, to my mind, undoubted plagiarisms. With regard to the behavior of Leibniz, when charged with these plagiarisms, Sloman is not beyond calling Leibniz a liar point-blank; I prefer to call his statements perversions of the truth, made under stress of circumstances, so that *his reputation as a great and original thinker should not suffer.* For instance, to explain what I mean, I will take the statement of Leibniz to de l'Hospital that he owed nothing to Barrow. As I have said in another place, from one point of view, the point of view that Leibniz would take for the *purposes of this letter,* Barrow would be a hindrance rather than a help to Leibniz, in the formulation of his *algebraical* calculus, *after he had once absorbed all the fundamental ideas.* That is, it would seem that Leibniz always tries to tell the truth, but to put it in a form that to the uninformed reader will convey quite a wrong impression. Another example of this juggling with words and phrases is given by Sloman, in the shape of a letter from Leibniz, dated August 27, 1676, *and the first draft of the same*; these two read together are very much the same, but read apart convey a totally different impression.

A second characteristic of Leibniz may also be traced back to his desire to make up for his lateness in starting; that is, the sometimes ridiculous claims that Leibniz makes to discoveries, or rather hints at having made them. An instance is given in the *Historia* (see above, Chapter III, p. 44). "It is required to form the sum of all the ordinates $\sqrt{(1-xx)} = y$; suppose $y = \pm 1 \mp xz$, from which $x = 2z/(1+zz)$, and $y = (\pm zz \mp 1)/(zz+1)$; *and thus again all that remains to be done is the summation* of rationals." Unless we assume that Leibniz never understood in all his life what we now call the change of the variable in integration, which to me seems rather far-fetched, the only reason why this should have been allowed to appear in a tract that was certainly written after 1712, is that Leibniz had never attempted this summation; he had set this down in 1674 and 1675 as a method of quadrature for the circle, not at that time having perceived the importance of the factor dz, or, in other words, the way in which the ordinates should be ordinated; for as I have already pointed out, at that time Leibniz could not have found dz, since he could not differentiate a product. This goes to prove that his reading of Pascal was not of the profoundest; for Pascal is very careful over this point, going to the trouble of calling the y's ordinates when drawn through the points of equal division of the base, and sines when they are drawn through the points of equal division of the arc. Probably to this characteristic is due the claim, set in italics in the manuscript above, with respect to equations of higher degrees. *He thought* he had a general method, which he had not time to verify by particular examples, and so find that his claim was erroneous. For surely this cannot be read as a claim to the Tschirnhausian transformation and the expression of a quintic in the canonical form $x^5 + px + q = 0$.

The date of the above manuscript is almost certainly antecedent to the manuscript that Leibniz got ready for the press, *De Quadratura*; hence his claim to be able to give examples of the calculus, except for integral powers which had already been done by Wallis, is without foundation.

With regard to the arithmetical quadrature itself, the great importance of it in the estimation of Leibniz is apparently in the correction and enlargement of tables; this claim, as Leibniz puts it, is ridiculous, although it could be so used by first constructing tables for angles whose tangents are given. But Leibniz, after giving a calculation true to twelve places of decimals, states that "the ratio of the circumference to the radius is now known," and proposes to use that. Apparently he does not see that to calculate this ratio from the series he gives, it will be necessary to take a billion or so of terms! For he does not give any hint of any modification of the series, or the use of the value obtained for some small angle.

Lastly, with regard to the calculation, it is strange that the denominator chosen as a common denominator is 15 times what it need have been; also it is a matter of wonder, considering that tables of logarithms were known to Leibniz, as a reader of Mercator and others, that Leibniz puts the matter in fractional form instead of working in decimals; thus, the arc whose tangent is 0.1 is equal to

$$
\begin{array}{ll}
0.1 & -0.00033333333333333\ldots \\
0.000002 & 0.00000001428571428\ldots \\
0.0000000001111\ldots & 0.000000000000909\ldots \\
\hline
=0.100002000111111\ldots & -0.000333347619956\ldots
\end{array}
$$
$$=0.99966865249\ldots\ldots$$

Finally, note that while Wallis and Brouncker are mentioned, Barrow is not. This is all part and parcel of his successful attempt to conceal, from all but Oldenburg, the fact that he had a copy of Barrow in his possession, right from the commencement of his studies.

III.

Transcribed from a manuscript tract of Newton on "Analysis by means of equations with an infinite number of terms."

AB⊓x, BD⊓y, a, b, c given quantities, m, n whole numbers. If then

$$ax^{\frac{m}{n}} \sqcap y, \quad \frac{an}{m+n}x^{\frac{m+n}{n}} \sqcap [\sqrt{\,} y] \sqcap \text{ area of ABD.}$$

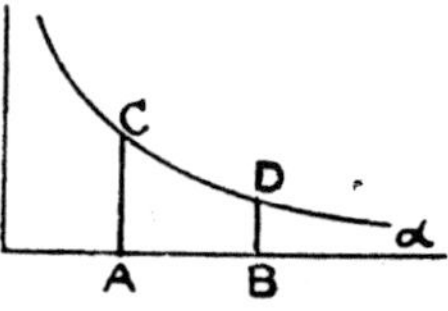

In connection with this the following example is to be noted:

If $\dfrac{1}{x^2}$ ($\sqcap x^{-2}$) $\sqcap y$, that is to say, if $a=1$, $n=-1$, and $m=-2$, then we shall have

$$\left(\frac{1}{-1}x^{\frac{-1}{1}}\sqcap\right)-x^{-1}\left(\text{or }\frac{-1}{x}\right)\sqcap a\mathrm{BD},$$

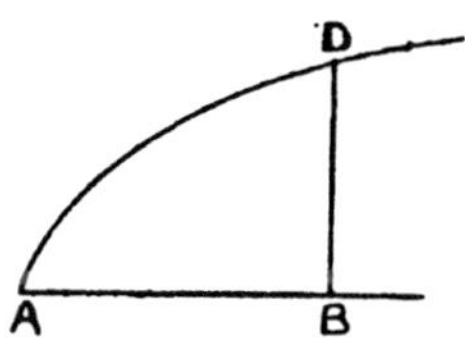

produced indefinitely in the direction of a; the calculation makes this negative because it lies on the other side of BD.

Again, if $\dfrac{1}{x}$ (or x^{-1}) $\sqcap y$, then $\dfrac{1}{0}x^{\frac{0}{1}}\sqcap\dfrac{1}{0}x^0\sqcap\dfrac{1}{0}x^1$ ($*$ this

ought to be written $\dfrac{1}{0}$ $1*$) $\sqcap\dfrac{1}{0}\sqcap$ infinity, which is the area of the hyperbola on either side.

If $\dfrac{1}{1+x^2}\sqcap y$, on division we obtain

$$y\sqcap 1-x^2+x^4-x^6+\text{etc., and then}$$

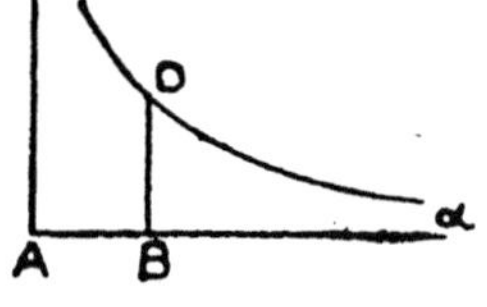

$$\mathrm{ABCD}\ \sqcap\ \frac{x}{1}-\frac{x^3}{3}+\frac{x^5}{5}-\frac{x^7}{7}+\text{ etc. ; or, if}$$

the term x^2 is the first in the division, the value of y will be $x^{-2}-x^{-4}+x^{-6}-$ etc.,

and hence BD $a\ \sqcap\ -\dfrac{x^{-1}}{1}+\dfrac{x^{-3}}{3}-\dfrac{x^{-5}}{5}+$ etc.

The first method is to be used when x is small enough, and the second when x is large enough.

Gerhardt then remarks that Leibniz has noted completely the following two cases of extractions of roots:

$$\sqrt{(a^2+x^2)}\ \sqcap\ y,\text{ and}\frac{\sqrt{(1+axx)}}{\sqrt{(1-bxx)}}\ \sqcap\ y.$$

Gerhardt further notifies the reader that he has omitted everything that he has found Leibniz to have copied out word for word, on comparison with Biot's edition of the *Commercium Epistolicum* (1856).

In the above, Leibniz marks interpolated remarks of his own with either [ ] or ($*$ $*$).

In the same manner, Leibniz has written out word for word the part of the manuscript dealing with the solution of adfected equations (against this he has put the final observation: "And these things that have been given will be sufficient for the investigation of areas of curves"), in addition to the part which follows, "the application of what has been given to other problems of the same kind," which, as being already known to him, he has not copied out. He goes straight on to the next section, "To find the converse of the foregoing, that is, to find the base when given the area, and to find the base when given the length of the curve." He has written this out word for word; also he has noted fully to the end the "proof of the method of solution of adfected equations."

At the end of these extracts from Newton's tract follow the words, "I extracted these things from the letter of Newton 20 Aug. to Newton." Gerhardt states that he has already said all that is necessary about the contents of these extracts.

SECOND SHEET.

Extracts from the correspondence between Collins and Gregory.

Among a number of partly illegible and unintelligible notes the following were to be noticed.

Gregory, January, 1670: Barrow shows himself to be most subtle in the geometry of optics. I think that he is superior to all whose works I have looked into, and I esteem this author beyond anything that can be imagined.

Sept., 1670: I think that Barrow has gone infinitely further than all those who have written before him. From his method of drawing tangents, combined with certain meditations of my own, I found a general geometrical method of drawing tangents, without calculation, to all curves, which not only contain his particular methods, but the general method as well. This is shown in 12 propositions.

Letter of Newton, 1672: ABC is any angle, $AB \sqcap x$, $BC \sqcap y$. Take, for example, the equation,

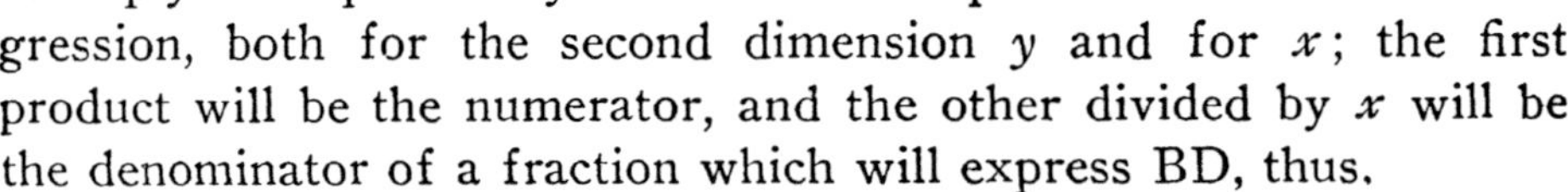

$$\overset{0}{x^3} \overset{1}{-2\,x^2 y} + \overset{0}{bx^2} \overset{0}{-b^2 x} \overset{2}{-by^2} \overset{3}{-y^3} \sqcap 0.$$

Multiply the equation by an arithmetical progression, both for the second dimension y and for x; the first product will be the numerator, and the other divided by x will be the denominator of a fraction which will express BD, thus.

$$BD \;\sqcap\; \frac{-2\,x^2 y + 2\,by^2 - 3\,y^3}{3\,x^2 - 4\,xy + 2\,bx - b^2}.$$

Moreover that this is only a corollary or a case of a general method for both mechanical and geometrical lines, whether the curve is referred to a straight line, or to another curve, without the trouble of calculation, and other abstruse problems about curves, etc. This method differs from that of Hudde and also from that of Sluse, in that it is not necessary to eliminate irrationals.

NOTE.

It is almost useless trying to write a critical note on the above in such an incomplete state. But I may remark that Leibniz apparently was at the time quite ignorant of what we now term "putting in the limits for a *definite* integral."

Gerhardt considers that the existence of this extract proves conclusively that Leibniz did not see the letter of Newton so often referred to; forgetting, as Sloman remarks, that Leibniz ought not to have seen the tract at all!

P. S. In allusion to Notes 3 and 18 on pp. 159 and 165, and Note 31 on p. 16, with regard to the use of the word "moment" or "momentum" as it is applied by Leibniz, I have found since they were written that Cavalieri, in his *Exercitationes Sex,* defines the term in the mechanical sense and gives much of the matter of Pascal on Centers of Gravity, as it appears in the "Letters of Dettonville." I suggest that Leibniz saw it in Cavalieri, and that its origin is to be traced to Galileo. See Note 25 on p. 208, where a precis of the fifth Exercitation is given. Duhem, in *Les origines de la statique* (Vol. I, pp. 134-144) attributes the first notion of a "moment" to an unknown, who is referred to as "the precursor of Da Vinci" (? Jordanus Nemorarius; cf. *Jordani opusculum de ponderositate Nicolai Tartaleae studio correctum,* etc. MDLXV).

The mechanical use of "moment" must be distinguished from the primary meaning of an inherent *force*: we see that in 1684 this still persisted. For, in the *Acta Eruditorum,* 1684, p. 511, there are extracts from the *De momentis gravium* of J. F. Vanni, 1684, in which the equilibrium of a sphere on two planes is considered; and here the "total moment" of the sphere is its weight, i. e., the *vis descensiva* of the predecessors of Cavalieri. Cavalieri however points out that the same body may have different moments for different positions.

VII.

LEIBNIZ AND PASCAL.[1]

(BY C. I. GERHARDT.)

IN the History of Mathematics it is generally stated that
the higher analysis took its rise in the method of indivisibles of Cavalieri (1635).[2] This assertion, at least as far
as the invention[3] of the algorithm of the higher analysis is
concerned, is erroneous. In what follows it will be shown,
by argument founded on the work of the French mathematicians of the seventeenth century and on the manuscripts
of Leibniz, that Leibniz was led to his invention of the
algorithm of the higher analysis by a study of the writings
of Pascal, more than by anything else.[4]

[1] [Translated from Dr. C. I. Gerhardt's article, "Leibniz and Pascal," in
the *Sitzungsberichte der Königlich Preussischen Akademie der Wissenschaften
zu Berlin*, 1891 (Zweiter Halbband), pp. 1053-1068. My own notes are put in
square brackets, to distinguish them from those given by Gerhardt.]

[2] "When I speak of the geometry of indivisibles," says Leibniz, "I intend
something far more comprehensive than the geometry of Cavalieri, which does
not appear to me to be anything but an insignificant (*mediocris*) part of the
geometry of Archimedes." [The general statement appears to me to be nearer
the truth than that of Gerhardt, who lays unjustifiable stress on the above
remark of Leibniz. I have endeavored to show later that there is strong probability that the work of Cavalieri, which Leibniz in the *Historia* acknowledges
to have read, was the *Exercitationes Sex*, and not the *Geometria* that was published ten years earlier; perhaps he read them both.]

[3] [It seems to me that those who claim merely the symbolism of the
Calculus as an "invention" of Leibniz are really detractors from his genius.
I have endeavored to show in the chapters previous to this, that this symbolism, more especially as regards the sign of differentiation, was a gradual
adaptation and development of ideas already preconceived for finite differences,
until Leibniz had obtained a standardized symbolism for the infinitesimal calculus. This, in my opinion, evidences an immensely greater intellect than that
necessary for an "invention"; even if we do take the standpoint that he was
helped by the work of his immediate predecessors. Perhaps Gerhardt's word
Erfindung might be better rendered by "construction" instead of "invention"
or "discovery."]

[4] [There was absolutely nothing in Pascal to suggest the sign or the rules
for *differentiation*, and Leibniz might just as easily have obtained his ideas on
integration from Galileo or others as from Pascal.]

With regard to the manuscripts of Leibniz, the first letters of the correspondence between Leibniz and Tschirnhaus are weighty; they contain the further discussion of their joint labor during the time that they lived together in Paris (September, 1675, to November, 1676);[5] it is well known that it was during this time that Leibniz invented the algorithm of the higher analysis. Among these letters, one from Leibniz, not hitherto published, which closes the first part of the correspondence between Leibniz and Tschirnhaus, contains a very detailed statement of the studies of Leibniz during his sojourn in Paris; it is beyond dispute of the utmost importance, since it was written only four years afterward and recalls particulars in a most vivid manner.[6]

Next, we have to consider the works of the French mathematicians about the middle of the seventeenth century, especially those of Pascal. We know from the facts of Pascal's life that his father, when he moved to Paris in 1631, joined a circle of mathematicians and physicists,[7] of which the history of science has preserved the names of Mersenne, Roberval, Gassendi, Desargues, de Carcavi, Beaugrand, des Billettes, and others. These were in com-

[5] [According to the generally accepted account, Leibniz was in London at the end of the third week of October, 1676, on his way home, via Amsterdam. At that time he *could not differentiate a product*.]

[6] [A point therefore to be carefully noticed is that the figure given for the characteristic triangle is totally different from that given in the "Bernoulli postscript"; it is also different from the figure used by him in the manuscript dated Oct., 1674, which is undoubtedly derived from the figure used by Pascal in the opening lemma to the *Traitté des Sinus du quart de Cercle* (compare the figures given on pp. 62 and 15); it is different from either of the figures used in the manuscripts of Oct. 29, Nov. 11, 1675 (see above, Chapter IV, pp. 78, 83, 102), the last of these being like Barrow's Differential Triangle, as used by him throughout his theorems on quadratures. Does this point to a new supposition: namely, that Leibniz originally invented a certain characteristic triangle of his own, essentially different in small detail from that of Pascal, Barrow, or any one else; that then he gradually passed from this to that of Pascal, later to Barrow's form; that he found this the most convenient of all; finally, through lack of memory, he ascribes the earliest form to Pascal, instead of to himself, making an erroneous apperception of the time at which he had *discovered* this early form? The point is referred to in a later note (40).]

[7] It went by the name of "Compagnie"; out of it grew, in 1666, the "Académie des Sciences."

munication, chiefly through Mersenne, with the mathematicians who did not live in Paris, Descartes, Fermat, and de Sluse; so that about the middle of the seventeenth century all that was best (*die Höhe der*) in the science of mathematics was concentrated in Paris.[8] In this circle Pascal moved, hardly yet out of his boyhood, and excited by his eminent talent astonishment and admiration. As an outstanding characteristic of the works of the mathematicians named above there stood forth the endeavor to abandon the method of Cavalieri as lacking every feature of scientific rigor, and to treat the science according to the methods of the Greek mathematicians.[9] Perhaps the ideas of Kepler, in his *Supplementum Stereometriae Archimedeae*,[10] were of influence, when Roberval and Pascal introduced into geometry the ideas of infinity and the infinitely small.[11]

As for those works of Pascal, which belong to this subject, we must mention in particular the solution of the problems, produced by him in 1658 under the assumed name of Dettonville, on the cycloid. By this, and by the method that he employed, he surpassed all the mathematicians contemporary with him, and he earned for himself the fame of being the greatest geometer of his day.

The investigation of the properties of the cycloid had occupied the attention of the most famous mathematicians

[8][Gerhardt no doubt here refers to French mathematicians; but the first-mentioned names, of those that lived in Paris, with the exception of Roberval hardly bear comparison with those of the three who did not live there.]

[9] The writings of Roberval and Pascal bearing reference to this have been mentioned in the essay "Leibniz in London." [Omitted in Chapter VI.]

[10] *Nova Stereometria Doliorum Vinariorum, inprimis Austriaci, figurae omnium aptissimae, et Usus in eo Virgae Cubicae compendiosissimus et plane singularis. Accessit Epitome Stereometriae Archimedeae Supplementum. Lincii an. M DC XV.* See my *Geschichte der Mathematik in Deutschland*, pp. 109ff.

[11] Roberval in a letter to the astronomer Hevelke (Hevelius) in Dantzig, writes: "Concerning analysis, in which I delight, I have far more [theorems]; and no fewer concerning the doctrine of the infinite, which they now call the 'doctrine of indivisibles'...." Published in: *Huygens et Roberval. Documents nouveaux. Par C. Henry*; (Leyden, 1879).

of the seventeenth century. It is reported that, earlier than anybody else and indeed before 1599, Galileo had had his attention called to this curve in consequence of his construction of arches for a bridge; he endeavored to find its area in a mechanical way, by weighing a plate of lead of uniform thickness having the shape of a plane bounded by a cycloid; and he found that it was very nearly three times as great as the area of the generating circle. This result he was unable to confirm theoretically. In 1615, Mersenne had his attention called to the cycloid as generated by a rolling wheel; he spent a great deal of time in investigating the nature of the curve, but without success; so that, in 1643, he corresponded with Roberval concerning the difficulties that he had encountered with respect to the curve. Roberval proved, by the help of the method of Cavalieri as improved by himself, that the area of the cycloid is exactly three times that of the generating circle; furthermore, in 1644, he determined the content of the solids formed by the rotation of the cycloid about its base, about its axis, and about the diameter of the generating circle; also he found the centroid of the area of the cycloid. In consequence of a bodily infirmity that robbed him of his rest at night, Pascal, in order to obtain some distraction from his pain, once more took up the investigation of the cycloid after an interval of fourteen years, in the year 1658. His design was to find the area of any chosen segment of the cycloid, the centroid of such a segment, the volumes of the solids described by such a segment by a rotation round either the ordinate or the abscissa, either by a complete, or a half, or a quarter revolution.[12] Inasmuch as the solutions of the problems hitherto investigated had not been done by any general method, but rather by special arti-

[12] [By ordinate and abscissa, Gerhardt means what Pascal calls the axis and base of the segment. Pascal only considered the whole solid of revolution, and the semi-solid, their volumes, their centers of mass, and the centroids of their surfaces; but those for solids generated by a quarter of a revolution could have been deduced quite easily.]

ficial ways of procedure, the question was that of specially
creating a treatment that was applicable in general. Pascal
reverted to the method of Archimedes, for determining the
quadrature of the parabola by means of the equilibrium of
the lever; he generalized the method,[13] by supposing, instead of geometrical figures, unequal weights not merely
at the extremities of the lever (which he follows Archimedes in terming *balance*) but also at several different distances from the fulcrum; of these, by means of the Arithmetical Triangle which he had invented,[14] he determined
the sum and the center of gravity.   On the advice of his
friends, Pascal, in June, 1658, under the alias of Dettonville,[15] determined to propose to mathematicians for solution the problems that he had solved.   October 1, 1658, was
settled as the last day for sending in solutions.   Particular
cases of the proposed problems were solved by Huygens,
de Sluse, and Wren, before the appointed day; but this
was not sufficient to meet the requirements of Pascal.   At
the request of de Carcavi, Pascal made known the above-
mentioned method for solving such propositions in a long
letter, at the beginning of October, 1658,[16] and added

[13][Pascal, in effect, obtained the general formula

$$\overline{x} = \Sigma(mx)/\Sigma(x),$$

where Σ stands for either a summation of finite quantities, or for the equivalent of integration.   If this is to be ascribed to Pascal as an original contribution, then we must assume that he had never seen Cavalieri's *Exercitationes
Sex, Exer. quinta*, Theorems 6, 7, 8, and certain others of the fifty propositions
that form this section of the book; the section being entirely devoted to centers
of gravity, while the method is a direct anticipation of Pascal's.]

[14][What is generally known as the Arithmetical Triangle is not mentioned in the *Lettres de Dettonville*; see Note 19, p. 204.]

[15][It may be of interest to note that the pseudonym of Amos Dettonville
is an anagram on Lovis de Montalte; Lovis, or Louis de Montalte being the
pseudonym under which Pascal's *Lettres provinciales* appeared.]

[16] Pascal published what he had written to de Carcavi along with the five
essays in the following year, under the title of: *Lettres de A. Dettonville contenant quelques unes de ses Inventions de Geometrie. Scavoir, La Resolution
de tous les problemes, touchant la Roulette qu'il avoit proposez publiquement
au mois de Juin, 1658. L'Egalité entre les Lignes courbes de toutes sortes de
Roulettes et des Lignes Elliptiques. L'Egalité entre les Lignes Spirales et
Paraboliques, demonstrée à la maniere des Anciens.   La Dimension d'un*

thereto three further propositions with respect to the cycloid. In this letter are combined five essays, which prepare the way for the solution of the problems of Pascal.

Solide formé par le moyen d'une Spirale autour d'un Cone. La Dimension et le Centre de Gravité des Triangles Cylindriques. La Dimension et le Centre de Gravité de l'Escalier. Un Traitté des Trilignes et leurs Onglets. Un Traitté des Sinus et des Arcs de Cercle. Un Traitté des Solides Circulaires. A Paris, M DC LIX. This writing contains the essays of Pascal of the year 1658 together with communications to Huygens, de Sluse, and an unnamed correspondent. From the correspondence of Huygens in the years 1658 and 1659, which is printed in that truly great work: *Oeuvres Complètes de Christiaan Huygens publiées de la Société Hollandaise des Sciences,* we see that a great movement arose among contemporary mathematicians through Pascal's problems, as well as through the printed works that we have mentioned. Leibniz expresses himself thus: "By this time, the controversy [referring to Gregory St. Vincent] had cooled down; when lo! fresh movements in the realm of geometry are stirred up through the whole of France, by Blaise Pascal, a man of the highest genius, and one who at that time had come nearer to the reputation of Galileo and Descartes than any one else."—This writing of Pascal was recommended for study to Leibniz by Huygens.

[As given by Pascal in his letter to de Carcavi, containing the particulars of his method for centers of gravity and the definitions of "trilignes" and "onglets," the problems proposed in June were:

1. To find the dimension and the center of gravity of the space CYZ.

2. To find the dimension and the center of gravity of its semi-solid of rotation about the base ZY, i. e., the solid formed by the triligne CYZ when rotated about the base ZY through half a turn only.

3. To find the dimension and the center of gravity of the solid of revolution about the axis CZ.

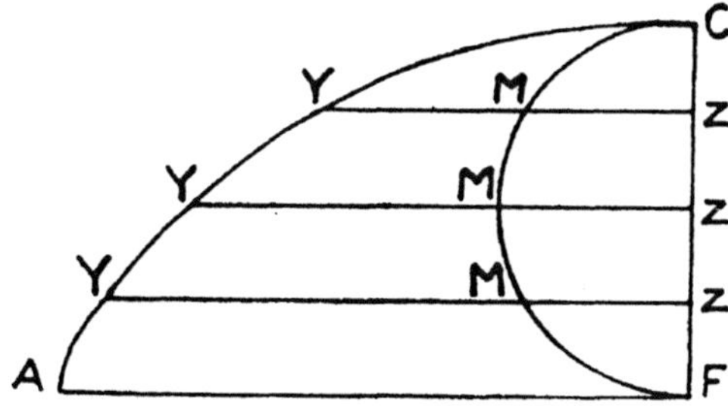

To which are added the three proposed in the *Histoire de la Roulette* at the commencement of October:

1. To find the dimension and the center of gravity of the curved line CY.

2. To find the dimension and the center of gravity of the surface of the semi-solid about the base.

3. To find the dimension and the center of gravity of the surface of the semi-solid about the axis.]

i. *Traitté des Trilignes et leurs Onglets.*[17]

In this essay, the determination of the content and the centroid of a "triligne" and its "double onglet" is reduced to the sum of the ordinates of the axis or the base in a triligne; also Pascal showed that the determination of the content and the center of gravity of the curved surface of the double onglet could be expressed as the sum of the sines of the axis.[18]

[17] By "Triligne" Pascal intends a plane figure bounded by two straight lines perpendicular to one another and a curved line. One of these perpendicular lines is called the axis and the other the base of the figure. If upon such a figure as a base there is erected a right solid, and this solid is cut by a plane which passes through the axis, or the base, then the portion of the

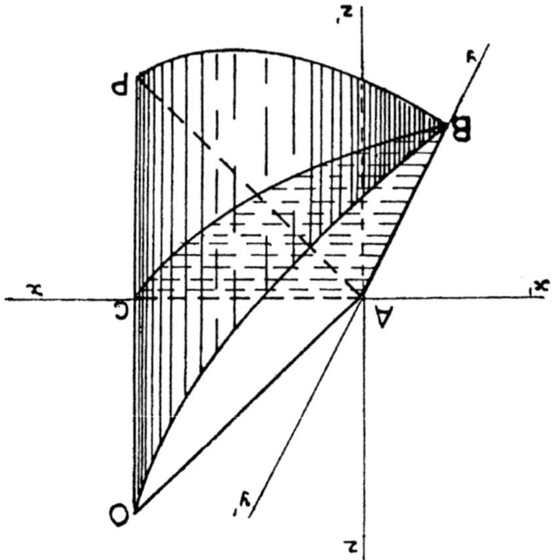

solid that is cut off is called an "onglet." A "double onglet" is obtained if, through the solid formed by production on the other side of the base, there is drawn a plane with the same inclination. [The last sentence does not make it clear that the second cutting plane also passes through the axis, or the base, as the case may be; nor that the plane is anticlinic and not parallel to the first plane; nor that Pascal took in general the inclination of the planes to the plane of the triligne to be 45°. I have therefore tried to represent the onglet and the double onglet in a diagram, see above.

ABC is the triligne, OABC is the onglet of (the axis or base) AB, and OBPCA is the double onglet of AB; the angles OAC, PAC are half right angles.]

[18] By *Sinus* Pascal intends the ordinates multiplied by the indefinitely small portions of the arc. [This is a very misleading statement; for Pascal especially distinguishes between *sines* and *ordinates,* and thus makes a considerable advance over his contemporaries. He defines them at the same time for

The next essay,

ii. *Propriétés des sommes simples, triangulaires et pyra-midales,*

is an appendix to the foregoing. By triangular sum, Pascal meant the sum of a number of magnitudes, each one multiplied in succession by the corresponding number in the natural scale. In the same way, a pyramidal sum de-finite section and for infinitesimal section; the distinction is made perfectly obvious in a diagram if we use finite section, say, division into four equal parts, of the quadrant of a circle as a special case of a triligne. Now the sum of the

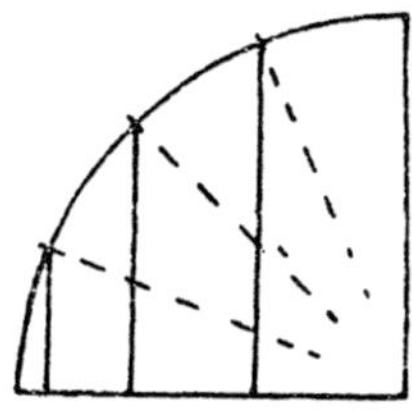

SINES OF THE BASE.

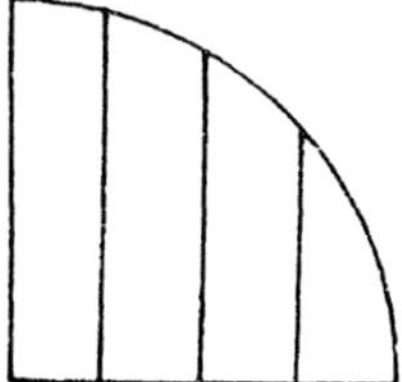

ORDINATES OF THE BASE.

sines or the ordinates are defined as the sum of the rectangles (for, as with all cases of indivisibles, that is what it comes to), formed by the sines or the ordinates respectively multiplied by the corresponding *equal* sectional parts. Thus, to speak of the sum of the sines as being the ordinates multiplied by the small portions of the arcs is quite wrong. Though only in rare cases is the space drawn, Pascal's idea of the sum of the sines is that of the space formed by *straightening* the arc and erecting at each point of division the correspond-ing sine. Now, as Pascal says in Prop. 1 of the *Traitté des Trilignes,* the sum of the ordinates, which have to be applied to the base, makes the figure itself; while in Prop. 1 of the *Traitté des Sinus du quart de Cercle,* he shows that the sum of the sines (as a special case of the general theorem quoted in iii by Gerhardt *supra,* p. 534) of a quadrant is equal to the square on the radius. Thus, in modern notation,

$$\text{sum of sines} = \int_0^{\frac{\pi}{2}} r \sin\theta \, . \, d(r\theta) = r^2$$

$$\text{sum of ords.} = \int_0^{\frac{\pi}{2}} r \sin\theta \, . \, d(r \cos\theta) = -\tfrac{1}{4}\pi r^2 \, .$$

The concluding paragraph of the *Traitté des Solides Çirculaires* runs thus: "All these results arise from the fact that the straight lines OI are ordinates, that is to say that they are equally distant and proceed from equal divisions of the diameter; this brings it about that the simple sum of the ordinates is the same thing as the space intercepted between the extremes. But this is not true for the sines, since the distances between adjacent ones are not equal to one another, and thus the sum of the sines is not equal to the space intercepted between the extremes; *there must be no mistaken idea on this point.*" We find the same care taken by Barrow; but Tacquet breaks down in determining the surface of a cone through not understanding the necessity of this point, and in consequence condemns the method of indivisibles.]

noted the sum of a number of magnitudes, each one in succession multiplied by the corresponding triangular number.[19]

Then comes,

iii. *Traitté des sinus du quart de Cercle*.

In this, Pascal begins by proving the theorem: "The sum of the sines of any arc of a quadrant of a circle is equal to the product of the part of the base, intercepted between the extremities of the outside sines, multiplied by the radius of the circle." By the help of this theorem, he investigated the sum of the sines of a quadrant of a circle, their squares, their cubes, fourth and higher powers,[20] the sum of the rectangles of each sine of the base into its distance from the axis, the triangular and pyramidal sums of the sines of the base, and so on.

[19][The effect is as Gerhardt states, but these sums are differently *defined* by Pascal in his letter to de Carcavi. The triangular sum of the numbers or magnitudes A, B, C, D, starting with A, (which should be stated), is the sum of all of them, plus the sum of all of them except the first, A, plus the sum of all except the first two, A and B, and so on; this is represented by Pascal as in the margin, and he goes on to show that this is equal to the first taken once, the second twice, and so on. Thus defined, the reason why they are named triangular numbers is obvious. The pyramidal sum is similarly defined as the triangular sum of all, plus the triangular sum of all except the first, plus the triangular sum of all except the first two, and so on. As if there were built up a pyramid having the first triangular sum as its bottom layer, the second triangular sum as the next layer, and so on; thus defined, the origin of the name pyramidal is obvious. Pascal then shows that this is the sum of the quantities taken respectively once, three times, six times, and so on, according to the sequence of the triangular numbers. Then using the property that twice a triangular number diminished by its ordinal number is equal to the square of that ordinal (i. e., $n(n+1)-n = n^2$), he also shows that, if two such pyramidal sums of quantities are taken, and from one of them the bottom layer is removed (i. e., the first triangular sum), then the sum of the two is equal to the sum of the quantities respectively multiplied in succession by the squares of the natural numbers. There is no connection between this and what is usually known as the Arithmetical Triangle of Pascal.]

[20][Pascal simply states the results, as deduced, not from the theorem quoted by Gerhardt, but (together with the theorem quoted) from the preliminary lemma that the radius is to the sine as the hypotenuse of the infinitesimal triangle is to its base: in modern notation, $r : y = ds : dx$, or $r\,dx = y\,ds$, where y is a *sine* and not an *ordinate* in Pascal's sense. All the following theorems are particular cases of the formula $\int y^n\,ds = r . \int y^{n-1} dx$.]

The next essay,

iv. *Traitté des sinus et des arcs de Cercle,*

contains the determination of the sum of all the arcs of a circle measured from the vertex of a quadrant to any ordinate of the axis, the sum of their squares, or their cubes, the corresponding triangular and pyramidal sums, the simple and triangular sums of the sectors, the sum of the solids formed from every sector of a quadrant and the distance of its center of gravity from the base, and so on.

v. *Petit Traitté des solides circulaires.*

In this is investigated the position of the center of gravity of such bodies as are formed by the rotation of half a band of a circle about the axis or base, the sum of the fourth powers of the ordinates of the axis, of their cubes, the position of the center of gravity of the semisolid of revotution arising from a rotation about the axis, and so on.

These five essays conclude with:

Un Traitté general de la Roulette, contenant la Solution de tous les Problemes touchant la Roulette qu'il avoit proposez publiquement au mois de Juin 1658.

All these works of Pascal are strictly geometrical in treatment, after the manner of the geometry of the ancients; there is not to be found in them a trace of the method of dealing with geometrical problems introduced by Descartes.[21]

It is well known that Leibniz through his acquaintance with Huygens, who lived in Paris from 1666 to 1681, was

[21] Descartes had spoken disparagingly about Pascal's "Essay on the Conics." Perhaps Pascal's decided opposition to Descartes may be traced back to this. Pascal's niece, Marguerite, writes: "M. Pascal used to speak very little about science; however, when the occasion for doing so occurred, he would state his

encouraged to study higher mathematics. More especially, it was Huygens who advised him to read the writings of Pascal. Upon several occasions later, has Leibniz declared, in conformity with this, that he was led to the higher analysis by the study of the writings of Pascal, and thus made his discoveries; first, in the hitherto unpublished letter to Tschirnhaus, of the year 1679, the part of it that relates to our subject being given later; also in a letter to the Marquis de l'Hospital, in the year 1694; further, in a postscript to a letter to Jacob Bernoulli, in the year 1703; and lastly, in the essay, *Historia et Origo calculi differentialis,* written in the last years of his life.

Up to the present time, among the manuscripts of Leibniz there has been found one of great length, that bears the title: *Ex Dettonvillaeno* (?) *seu Pascalii Geometricis excerpta: cum additamentis.* It is not dated; but as it contains work that is in the closest connection with the writings of Pascal to de Carcavi, hence it must be assigned very approximately to the time of his intercourse with Huygens (1673). This cannot be given in its entirety; only the commencement of it follows under the heading III. One special remark has Leibniz made on the five essays which follow Pascal's letter to de Carcavi; he states that the method of Pascal for determining the surface of the

opinion on those matters about which people were speaking to him. For example, with reference to the philosophy of Descartes, he merely said what he thought. He was of the same opinion as Descartes concerning automatism, but far from being so on the "subtle matter," which he ridiculed. But he could not put up with his (Descartes's) method of explaining the formation of the universe, and he often said: "I cannot pardon Descartes. In the whole of his philosophy, he would have been highly pleased to have dispensed with God; but he could not help making use of him to give a fillip to set the universe in motion. That being done, he had no further use for God." (Fougère, *Lettres, Opuscules et Mémoires de Madame Perier et de Jacqueline, soeurs de Pascal, et de Marguerite Perier, sa nièce.* Paris, 1845, p. 458). [It is more probable that Pascal used geometry, as Barrow did, because he both preferred it and thought it more rigorous than analysis. With regard to the remark on method, Gerhardt does not intend to convey the impression that Pascal abandoned for the more strictly geometrical method of moments the mechanical idea of the *balance,* with which he commences. By the way, to the best of my belief, the word "moment" is never used by Pascal.]

sphere,[22] according to which the surface of a solid formed by the rotation round an axis can be reduced to a plane figure proportional to it, was what induced him to make out a general theorem applicable to all plane figures bounded by a curved line.

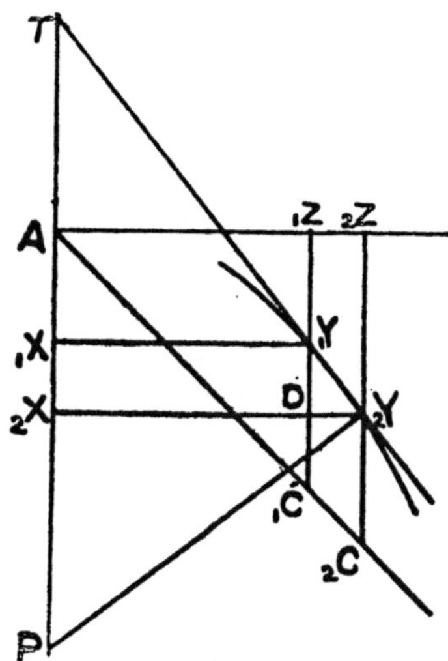

The coordinates of $_1Y$ and $_2Y$, two points on the curve, are $_1Y_1Z$, $_1Y_1X$ and $_2Y_2Z$, $_2Y_2X$; $_2YT$ is the tangent at $_2Y$, which is supposed to meet the curve again in $_1Y$, and the normal $_2YP$ is drawn. On account of the similarity of the triangles $_1YD_2Y$ and $_2Y_2XP$, we have

$$_2XP \cdot _1YD = _2Y_2X \cdot _2YD;$$

i. e., the subnormal $_2XP$ applied, at right angles to the axis AX, to the element of the axis $_1X_2X (=_1YD)$, is equal to the ordinate $_2Y_2X$, applied to the element $_2YD$.[23] "But,"

[22][I have gone carefully through the "Lettres of Dettonville," and I find no mention of Archimedes except in one place, namely, Prop. 1 of the *Traitté des Solides Circulaires*; and the whole of this is devoted to *volumes* of solids and their centers. Nor can I find any place where Pascal determines the surface of a sphere, at least not by reducing it to an equivalent plane figure, I have however shown that Barrow does do this (see above, Chapter III, p. 58). Surely Leibniz must be confusing the work of Pascal with that of Barrow on quadratures, the latter being so similar to the former in places that Barrow might easily be suspected of "borrowing" from Pascal; much more easily indeed than Leibniz could be so suspected with regard to either, in spite of his own assertion with regard to Pascal. See Notes 23, 24.]

[23][These are far more like Barrow's results than those of Pascal; while the style is entirely Barrovian and quite different from that of Pascal.]

Leibniz continues, "straight lines which increase from nothing, each multiplied by its corresponding element, form a triangle. For, let AZ be always equal to ZC, and you get the right-angled triangle AZC, which is half the square on AZ, and thus the figure produced by applying the sub-normals in order at right angles to the axis is always equal to half the square on the ordinate. Hence, being given a figure to be squared, that figure is sought whose sub-normals are equal to the ordinates of the given figure, and the second figure is the quadratrix of the given figure. Thus from this very simple idea, we have the reduction of surfaces produced by rotation to plane quadratures, and also of the rectification of curves;[24] and at the same time, we can reduce these quadratures to problems of inverse tangents." Thus it came about that Leibniz obtained from this a general method for the quadrature of curves.

All this was arrived at by Leibniz in the first year, 1673/74, of his mathematical studies in regard to the higher analysis. Until this time he had adhered to the rigorous geometrical method, as he found it in the writings of Pascal, in his investigations; acting on the advice of Huygens, he now made himself acquainted with the method of Descartes as being more adapted to computation. The long essay of Leibniz with the title, *Analysis Tetragonistica ex Centrobarycis*, dated Oct. 25, 26, 29, and Nov. 1, 1675, shows clear connection[25] with the above-mentioned method

---

[24] [There is no rectification of curves in Pascal; the whole of this sentence would however serve as a summary of the work of Barrow on rectification.]

[25] [Gerhardt states that the Centrobaryc Method, as considered by Leibniz in the manuscripts dated October 25, 26, 29, and November 1, 1675, shows clear connection with the work of Pascal. He asserts that, from a consideration of Archimedes, Pascal was enabled to extend the method of the ancients; he does not seem to be aware of what Cavalieri had done and published as the fifth section of his *Exercitationes Sex*; or else, knowing all about this, he suppresses that knowledge for fear of discrediting the statements of Leibniz concerning the methods of Cavalieri.

The striking points about the work of Cavalieri in question are as follows. He opens by defining gravity as a property of a body, a descensive force. He then defines a heavy body as one possessing this property, and in a note on the definition, he adds that these must be taken to include surfaces, lines, and

of Pascal; also it shows the improvement that Leibniz had made in consequence of his study of Cartesian geometry,

points. Then he gives the definition of "moment" in its mechanical sense. "The moment of a weight is its endeavor to descend, no matter at what distance it is hung." This is followed by the note: "Since this moment is different at different distances, as will be seen in what follows, it is to be understood from this that the same weight may have different moments." He then defines uniform and uniformly variable (*difformis*) weights, such as a parallelogram in which the density varies as some power of the distance from one side; also he defines the centers of gravity and equilibrium. In Prop. 6 he shows that the moments of bodies are compounded of the ratio of their weights and the ratio of their distances. In Prop. 8 et seq., he combines the doctrine of indivisibles with that of moments to find the centers of gravity of surfaces, chiefly by means of "analogous figures"; thus, a uniform triangle is analogous to a parallelogram whose "difformity is of the first species," i. e., the density varies as the distance from one edge. He shows that, if the difformity is of the nth species, i. e., if the density varies as the nth power of the distance from the edge, then the medial line is divided by the center of gravity into parts in the ratio of 1 to $n + 1$, although it is stated rather differently, and only worked out for the first few values; then, using the idea of moments he proceeds from one degree to another in the case of the triangle, where the axis of moments (*limes*) is a parallel to the base through the vertex, and in the following proposition, the base itself; next the semicircle and the hemisphere are dealt with, whether uniform or varying as the distance from the center. In Prop. 36, he lays down the idea that the axis of moments may be outside the figure under consideration; and then proceeds to consider cylinders, cones, parabolic conoids, and the sphere, and truncated portions of them; and finally he finds the moment of a portion of a hyperbola about the asymptote which is not the base of the portion considered. It is interesting to note that Cavalieri, when speaking of the difformity of weight, uses the phrase "*incrementum difforme gravitatis,*" i. e., the word *incrementum* is employed to connote a *gradual* increase that follows a definite law. Also it is worthy of remark that he employs the notation, *o. l., o. p., o. q., o. c.,* etc. for "all the lines," "all the planes," "all the squares," "all the cubes," etc.

From the above it will be seen that Cavalieri has given a fairly comprehensive account of the use of moments for the determination of the center of gravity; thus he not only gives far more than Pascal, but anticipates him. Leibniz's matter is far more like that of Cavalieri than that of Pascal; though he seems to be reading Pascal at the time he wrote the third part of the "Analytical Quadrature," by the method of moments, for the last figure in this manuscript (see above, Chapter IV, p. 89), with the explanatory diagram that I have added on the right of it, is strongly reminiscent of the idea of the onglet of Pascal; although it may have arisen from Cavalieri's work. The great point about this batch of manuscripts of October and November, 1675, is that nearly every figure has the tangent drawn to the curve; now the tangents are never drawn or used either by Cavalieri or by Pascal. A secondary consideration, but still one of importance, is that the subject-matter of these manuscripts is like nothing in Cavalieri or Pascal, as far as the "center of gravity method" is concerned. As we find Pascal's Infinitesimal Triangle idea in the figure of Leibniz's manuscript of October, 1674, I take it that this was the time at which he *finished reading* his Pascal. Hence, I imagine that in October, 1675, he had got a good knowledge of Descartes's algebraical geometry, and began to study Cavalieri's *Exercitationes Sex*; he did not get very far in this before he appreciated the power given by the method of moments; then, probably wearied by Cavalieri's prolix demonstrations, he laid the book aside, and applied Cartesian analysis to the method of moments, running the idea for all it was worth. If this is the case, these manuscripts represent real *original* research, and are not study notes like some of the others.]

Leibniz commences with Proposition 2 from Pascal's first essay, *Traitté des Trilignes et leurs Onglets,* which he expresses as follows.

"Let any curve AEC be referred to a right angle BAD; let AB⌐DC⌐a,[26] and let the last x⌐b; also let BC⌐AD ⌐y, and let the last y⌐c. Then it is plain that

$$\text{omn. } \overline{yx \text{ to } x} = \frac{b^2 c}{2} - \text{omn. } \overline{\frac{x^2}{2} \text{ to } y} .$$

For, the moment of the space ABCEA about AD is made up of rectangles contained by BC ($=y$) and AB ($=a$);[26] also the moment about AD of the space ADCEA, the complement of the former, is made up of the sum of the squares on DC halved ($=x^2/2$); and if this moment is taken away from the whole moment of the rectangle ABCD about AD, i. e., from[26] c into omn.x, or from[26] $b^2 c/2$, there will remain the moment of the space ABCEA.

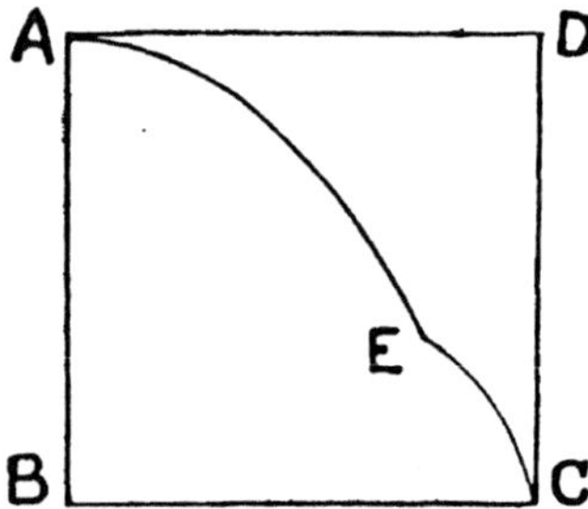

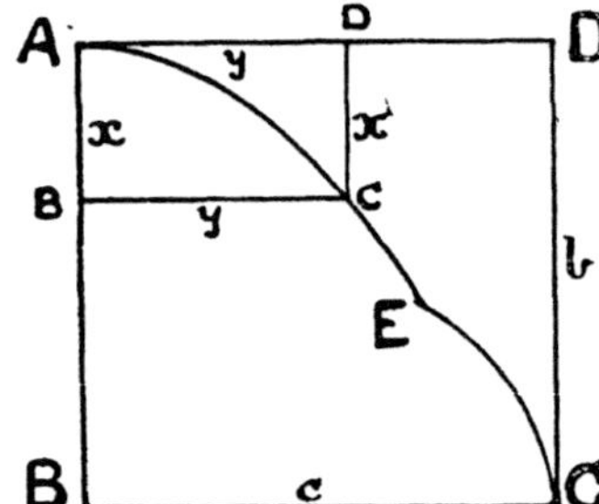

Hence the equation that I gave is obtained; and, by re-arranging it, it follows that,

$$\text{omn. } yx \text{ to } x + \text{omn. } x^2/2 \text{ to } y = b^2 c/2.$$

[26] [The misreadings of Gerhardt, as given in his *Geschichte der höheren Analysis* (see above, Chapter IV, p. 65) are uncorrected even in 1891, the date of this essay, thirty-six years after the publication of the *Geschichte*! We should have "AB⌐DC⌐x" and "AB($=x$)"—see the figure on the right (above) which is mine, while that on the left is the one that Gerhardt gives as that of Leibniz; again Gerhardt's "id est ac in omn.x, sive $a(cb^2/2)$," which makes Leibniz write nonsense, should be "id est a c in omn.x, sive a $cb^2/2$," the "a" being the preposition "away from" and not the length of a line; thus corrected we not only have a sensible reading but the whole paragraph is correct; I have made the correction when translating. Also with regard to Gerhardt's statement that Leibniz starts from an alternative rendering of Prop. 2 of Pascal's *Traitté des Trilignes,* it is worthy of remark that Pascal's

In this way we obtain the quadrature of the two joined in one in every case; and this is the fundamental theorem in the center of gravity method."

In the continuation, dated October 29, 1675, in connection with this theorem, Leibniz brings in the characteristic triangle, which has already been mentioned above.

AGL is any curve, $BL = y$,
$WL = l$, $BP = p$, $AB = x$,
$GW = a$, $y = \text{omn.}\, l$;

hence

$$\frac{l}{a} = \frac{p}{y} = \frac{p}{\text{omn.}\, l}\, ,$$

and therefore $p = \dfrac{\text{omn.}\, l}{a}\, .\, l\, .$

Now, by the theorem given above,[27]

$$\text{omn.}\, p = \frac{y^2}{2} = \frac{\overline{\text{omn.}\, l}\,^{\boxed{2}}}{2} = \frac{\overline{\text{omn.}\, l^2}}{2}\, ;$$

hence

$$\frac{\overline{\text{omn.}\, l^2}}{2} = \overline{\text{omn.}\, \overline{\text{omn.}\, l}\, .\frac{l}{a}}\, ;$$

"that is," adds Leibniz, "if all the l's are multiplied by their last, and all the other l's again are multiplied by their last, and so on as often as it can be done, the sum of all

figure is altogether different from that of Leibniz; and this is only natural, because there is *no similarity between the theorems, nor is there any relation between the methods of proof.* Pascal's proof is equivalent to the modern method of a change in the independent variable by a conversion to a double integral followed by a change in the order of integration, and is geometrical; that of Leibniz is equivalent to integration by parts, and is merely an example of the theorem of moments.

Thus (Pascal), $\int yx\, dx = \int(\int x\, dx)\, dy = \int \tfrac{1}{2}x^2 dy,$

and (Leibniz), $\int yx\, dx = [\tfrac{1}{2}x^2 y] - \int \tfrac{1}{2}x^2 dy;$

where Pascal's integrals are taken over the *same* area as one another, and those of Leibniz over *complementary* areas. It seems therefore ridiculous to say that "Leibniz commences with Prop. 2.... which he expresses as follows."]

[27][This means the result obtained geometrically by means of the triangle AZC, in the passage to which Note 23 refers.]

these will be equal to half the sum of the squares, of which the sides are the sums of these, or all the l's. This is a very fine theorem, and one that is not at all obvious. So is also the theorem,

$$\text{omn. } xl \sqcap x \,.\, \text{omn. } l - \text{omn.omn. } l,$$

where l is supposed to be a term of a progression, and x the number which expresses the position or ordinal that corresponds to the l, i. e., x is the ordinal number and l the ordered quantity.

N.B. In these calculations, a law for all things of the same kind may be observed; for, if 'omn.' is prefixed to a number or ratio, or to something indefinitely small,[28] then a line is produced, also if to a line, then a surface, or if to a surface, then a solid; and so on to infinity for higher dimensions.

It will be useful[29] to write $\int$ for 'omn.,' so that

$$\int l = \text{omn. } l, \text{ or the sum of all the } l\text{'s.}$$

Thus, $\int \dfrac{\overline{l^2}}{2} = \int \overline{\int \overline{l} \dfrac{l}{a}}$,   and $\int \overline{xl} = x\int \overline{l} - \int \int l .$ ''

This was the first time that the algorithm for the higher analysis was introduced. In what then follows, Leibniz obtains the first theorems of the integral calculus:

$$\int x = x^2/2, \ \int x^2 = x^3/3,$$

and adds, "All these theorems are true for series in which the differences of the terms bear to the terms themselves a ratio which is less than any assignable quantity."

Further Leibniz remarks: "These things are new and noteworthy, since they lead to a new kind of calculus. Being given l, and its relation to x, required to find $\int l$. Now this may be obtained by a reverse calculation; thus, if $\int l = ya,$

[28] [The connection between number, ratio, and *infinitesimal* is peculiar.]

[29] [Note the word "useful" (*utile*): the "long s" is introduced merely as a convenient abbreviation in accordance with Leibniz's usual idea of obtaining simplification by means of symbols.]

suppose that $l = ya/d$, that is to say, as $\int$ increases the dimensions, so d will diminish them; but $\int$ stands for a sum, and d for a difference.[30] From the given value of y, we can always find y/d or l, or the difference for the y's."

In the investigation that bears the title, *Methodi tangentium inversae exempla*, dated November 11, 1675, Leibniz introduces instead of y/d the notation dy.

Such are the chief points in the story of the introduction of the algorithm of the higher analysis, as far as may be gathered from the extant manuscripts of Leibniz.[31]

In connection with the earlier essay, "Leibniz in London,"[32] I have shown that any influence whatever from external sources upon Leibniz with regard to the introduction of the algorithm of the higher analysis is excluded.

[30] [I have discussed this fully in my translation of Gerhardt's essay, "Leibniz in London" (see above, Chapter VI, p. 180). I have shown there that at least it is highly probable that the d in x/d stands for a certain length, namely the subtangent.]

[31] [Note that, in spite of Gerhardt's opening remarks about the algorithm of the calculus being due to reading Pascal, the symbols of integration and differentiation have not been mentioned in anything quoted by Gerhardt in this essay, except in the paragraph just above.]

[32] [See critical notes on this point, Chapter VI, pp. 172-184. I believe some of those who read what is there given will, while giving Leibniz full credit for the introduction and *development* of the symbols $\int$ and d, that made the calculus of Leibniz the powerful instrument it was, still find it hard if not impossible to agree with Gerhardt in his assertion that the ideas of Leibniz were not very strongly influenced by the best points of every single author that he studied, and more especially by the *Lectiones Geometricae* of Barrow and the *Exercitationes Sex* of Cavalieri.]

TRANSLATIONS OF THE MANUSCRIPTS

Alluded to by Dr. Gerhardt.

I.

From the letters of Leibniz to Tschirnhaus.

1679.

"You are astonished that Reginaldus[33] should have been able to fall into error over the surface of an elliptic spheroid; but you do not seem to have considered sufficiently how different are the several methods of indivisibles. He certainly understands the Cavalierian method, but that is so circumscribed by narrow limitations that few things of any great importance can be obtained from it. There is no doubt that Cavalieri, Torricelli, Roberval, Fermat, and indeed, as far as I know, all the Italian mathematicians were quite unaware of the utility of tangents for the purpose of finding quadratures, or of that which I have been accustomed to call the infinitely small "characteristic triangle" of the figure; indeed, at the present time also in France, I believe that Huygens is the only man that really understands these matters.[34] Pascal himself could not sufficiently express his admiration for the artifice by which Huygens found the surface of the parabolic conoid. Sluse has given no example of these things, by which I am inclined to think that they are unknown to him also. This too is the reason why Huygens and Gregory demonstrated such theorems by roundabout methods, suppressing their analysis, in order not to divulge their method at once so easy and so fruitful.

[33] [So far I have failed to find any information as to the error into which Reginaldus fell; he does not appear to be mentioned by either Cantor or Zeuthen.]

[34] [The Geometry of Cavalieri is indeed practically all quadratures; but Torricelli himself says (quoted by Tommaso Bonaventura in his preface to an edition of the *Lezione Accademiche,* 1715), in his preface to a *Tract on Proportion,* that he has used indivisibles for tangents as well as for quadratures; Roberval, through his own efforts at concealing his methods, we know comparatively little about; but the germ of Fermat's method is the same as that of Barrow's, namely the Differential Triangle; lastly it is probable that Huygens's knowledge was considerably more than he let anybody know (and so too with Gregory)—cf. Leibniz's words, "suppressing their analysis," a few lines later. It is to be observed that Leibniz deliberately speaks of the mathematicians of France and Italy only; "at the present time," 1679, he must have been aware that Barrow had complete geometrical knowledge, at any rate, of *all* the matters in question.]

"The prime occasion from which arose my discovery of the method of the Characteristic Triangle, and other things of the same sort, happened at a time when I had studied geometry for not more than six months. Huygens, as soon as he had published his book on the pendulum, gave me a copy of it; and at that time I was quite ignorant of Cartesian algebra and also of the method of indivisibles,[35] indeed I did not know the correct definition of the center of gravity. For, when by chance I spoke of it to Huygens, I let him know that I thought that a straight line drawn through the center of gravity always cut a figure into two equal parts; since that clearly happened in the case of a square, or a circle, an ellipse, and other figures that have a center of magnitude, I imagined that it was the same for all other figures. Huygens laughed when he heard this, and told me that nothing was further from the truth. So I, excited by this stimulus, began to apply myself to the study of the more intricate geometry, although as a matter of fact I had not at that time really studied the Elements. But I found in practice that one could get on without a knowledge of the Elements, if only one was master of a few propositions. Huygens, who thought me a better geometer than I was, gave me to read[36] the letters of Pascal, published under the name of Dettonville; and from these I gathered the method of indivisibles and centers of gravity, that is to say the well-known methods of Cavalieri and Guldinus. I immediately committed to paper certain things that occurred to me as I read Pascal, of which I now find that some are absurd, others

[35] [The *Horologium* was published in March or April, 1673, and the presentation of a copy to Leibniz was undoubtedly made *after* his return from his first visit to London (Cantor says that the dedication was dated March 25, 1673; see Cantor, III, p. 138). Hence, the date at which Leibniz obtained the Characteristic Triangle can be assigned to some time at least not later than the beginning of October, 1673; and therefore the inclusion of this in the manuscript dated Aug., 1673 (see above, Chapter IV, p. 59), marks the exact date of its discovery.]

[36] [In the "Bernoulli postscript" (see p. 14) Leibniz states that he "sought a Dettonville from Buotius, a Gregory St. Vincent from the Royal Library, and started to study geometry in earnest." In the *Historia* (see p. 37) Leibniz says that "in order to obtain an insight into the geometry of quadratures, he consulted the *Synopsis Geometriae* of Honoratus Fabri, Gregory St. Vincent, and a little book by Dettonville (Pascal)." In his letter to the Marquis de l'Hospital he says, "At the start I only knew the indivisibles of Cavalieri, and the 'ductions' of Father Gregory St. Vincent, along with the 'Synopsis of Geometry' of Father Fabri" (see below, p. 220). I suggest that the correct explanation of these inconsistencies is that he did get the Dettonville from Huygens as stated here, the St. Vincent from the Royal Library, and the work that he obtained from Buotius was the *Exercitationes Sex* of Cavalieri.]

please me very much even at the present time.[37]   Amongst other things, I tried to find a new sort of center.   For, I thought that if, to any figure that was given, others that were similar and similarly placed were inscribed, then a "middle point" could be found,[38] at which the figure evanesced, and that being given this point the quadrature could be obtained; later I perceived the difficulty that

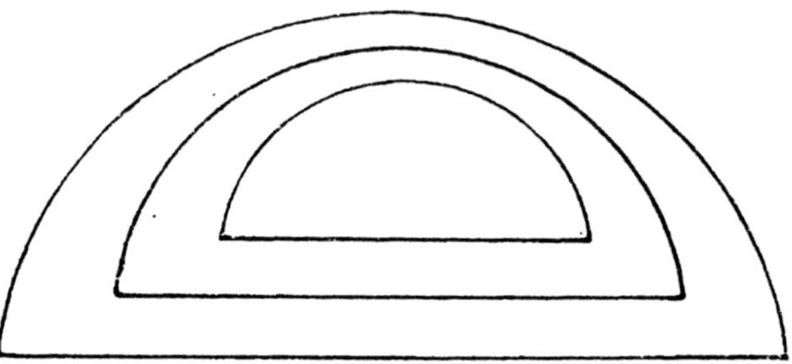

made this method ineffective.   But to return to the subject, I will tell you how I came to find the method of the Characteristic Triangle.   Incidentally Pascal gave a proof of the dimension of the spherical surface proved by Archimedes, that is the moment of a circular curve round the axis,[39] and showed that the radius applied to the axis produced this moment.   I, having examined the demonstration with care, observed that, with the aid of the infinitely small characteristic triangle, it was possible to prove the following general proposition for any curve:[40]

[37][I think the passage throws considerable light on the character of these manuscripts, besides explaining how it was that Leibniz seems to have taken a very long time to study the works of the authors mentioned.   I look on these manuscripts, not as "study notes" merely, nor yet as true "research," but as a mixture of each.   I suggest that there is quite enough evidence to make it safe to assert that the characteristic of Leibniz's method of study was to read a very small portion of an author at a time, then to break off and follow out the train of ideas suggested to him by the passage to the furthest limit, before proceeding further with his reading; thus he is led to his own *original* developments.   For instance, note in the next sentence how he says he "tried to find a *new* sort of center."   This is very characteristic; he is not satisfied with merely acquiring knowledge, even at this early stage, but at once seeks to utilize each point, *as he grasps it,* to obtain something new, something original previously undiscovered.   Cf. the study notes on the work of Pascal, given below under III.]

[38][That is, a "homothetic center."]

[39][As I have been unable to find the word "moment" defined, or even mentioned, in any place except in the *Exercitationes Sex* of Cavalieri, I suggest that this is fairly good circumstantial evidence for the reading of this work by Leibniz before he discovered the theorem in question.]

[40][Observe that this is not the figure used in the manuscript of October, 1674 (see above, Chapter IV, p. 62), the latter being a diagram that one would naturally expect him to have obtained from the figure in the lemma that

"Let AP be any curve and let BP be drawn perpendicular to
its tangent AT, to meet the axis in B; then, the ordinate PC being
drawn, let the straight line CD be applied to the axis AC, perpendicular to it, and equal to BP. Then if a curve is drawn through
all such points as D, we shall have a figure whose area will be the
moment of the original curve about the axis, i. e., it will show how

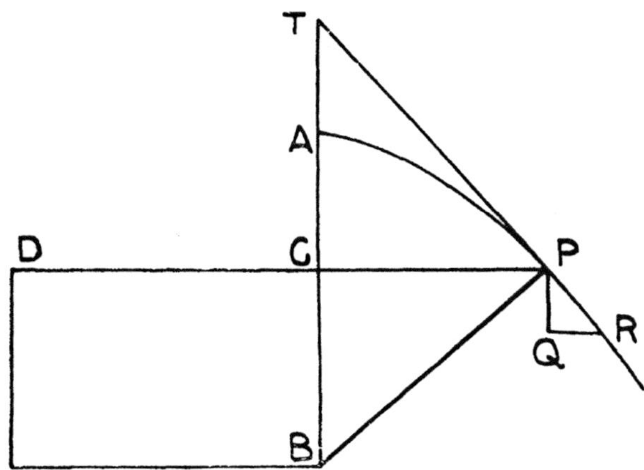

to draw a circle equal in area to that of the surface of a curve
rotated round the axis. Since in the circle the straight line BP
is always of the same length wherever the point P is taken in the
curve, hence the figure produced by the perpendiculars[41] applied to
the axis is a rectangle, and thus the surface of the sphere is very
easily reduced to a plane area. Now, when from this method I
had deduced a general method for the dimensions of such surfaces,
I at once took it to Huygens; he was surprised and laughingly
confessed that he had made use of precisely the same method for
obtaining the surface of the parabolic conoid of revolution. For
in that case the curve through every D is a parabola, and hence the

commences Pascal's *Traitté des Sinus du quart de Cercle* (cf. Note 6, p. 196);
but is a figure such as one would expect Leibniz to abstract from those given
by Barrow, either from Lect. XII, prop. 1, 2, 3, or from Lect. XI, prop. 1
(see Chapter IV, p. 58, and Chapter I, p. 16, respectively. In the latter especially
we have the right-angled triangle used by Leibniz on page 39, quoted by
Gerhardt in the article translated in the present number). I therefore suggest
that Leibniz worked at Barrow and Pascal conjointly, and applied Descartes's
analysis to their geometrical theorems. If this is not the case, Leibniz was at
fault, for Pascal was discussing *sines* and not *ordinates* (see Note 18, p. 202);
i. e., Pascal was integrating with regard to θ and not with regard to x. Observe
also that the figure as given is not correct; the rectangle should be that having
AC, CD as adjacent sides.]

[41][Note that the area is taken to be produced by the assemblage of lines
applied in order, in the true Cavalierian style.]

figure is capable of quadrature. Since I wished to verify the accuracy of my result in the case of the parabola,[42] I began to look for a method of expressing spaces and curves by reckoning, and then for the first time I really understood those matters of which Descartes wrote. For, previously, I used to calculate in my own way, using not letters but the names of lines. Then, for the first time, I read Descartes and Schooten carefully, acting on the advice of Huygens, who told me that the method of reckoning adopted by these authors was very convenient. Meanwhile having once opened the door provided by the characteristic triangle, I very easily discovered innumerable theorems with which at that time I filled innumerable sheets; but later I found that these had also been noted by Huraet, Gregory, and Barrow.[43] Moreover all these things I came upon in the first year of my apprenticeship to geometry. But after that I struggled forward to far greater things, such as I believe that neither Gregory nor Barrow could ever have reached by their methods, far less Cavalieri or Fermat.[44] About the same time, since I perceived that the finding of quadratures could be reduced to the finding of sums of series, and that the finding of tangents could be reduced to the finding of differences, I put together the

[42] [Query: urged thereto by a question on the part of Huygens, as to whether Leibniz could now find the properties of the auxiliary curve (see p. 18).]

[43] [This fits in perfectly with my suggestion that Leibniz attacked Barrow's *Lectiones* at several different times. Having, as I think, taken Barrow's advice given in the preface, he sampled the first few propositions of each lecture, and obtained from those of Lect. XI and XII his Characteristic Triangle. This could I think have been definitely settled if Gerhardt had only given the figure used by Leibniz in the manuscript dated August, 1673. Assuming for the time being that my suggestion is correct and that Leibniz is merely confusing the author that he read at this time, I suggest that characteristically he broke off his reading of Barrow, pursued the idea he had obtained, and made out those theorems on quadratures that he speaks of; this so improved his geometry that later he was able to read Barrow thoroughly and appreciate all that was in it, and to find that his theorems had been anticipated. I also suggest that it was on this second or third reading that he came across the theorem that led to his Arithmetical Tetragonism. A fresh reference to Barrow to find if there were any other ideas that he could develop, considerably later, having already found him a mine of information, would then probably be the occasion on which the marginal notes in his own notation were inserted by Leibniz.]

[44] [Leibniz seems to have got these men in true perspective, Cavalieri, Fermat, Gregory, and Barrow, as far as the infinitesimal calculus is concerned. But I doubt whether he, even after he came to his fullest appreciation of Barrow's *geometrical theorems,* or indeed any other person except Bernoulli, ever appreciated the real inwardness of these theorems, or that Barrow's tangent problems could be used, in the manner I have shown in the appendix to my Barrow, to draw a tangent to any curve *given by an equation in either Cartesian or polar coordinates.*]

fundamental principles of my new calculus,[45] which I call the "differential or tetragonistic calculus," by which I can set with a few little lines those things which could be obtained with great difficulty, if indeed at all, by the help of a mighty apparatus of lines. Moreover I considered in general that the finding of the sum of any series was nothing else but the discovering of some other series, the differences of the terms of which gave the given series, and this other series I used to call the summatrix.[46] The occasion for considering infinite series arose from the work of Wallis and Mercator.[47] When I joined their discoveries to mine, I found out new things with no trouble at all.

"At length, when I considered that problems of quadratures might not be of known degree, and yet might be reduced to equations, in which the exponents of the powers were unknowns, a new light dawned upon me and I began to understand that this was something beyond the ordinary analysis, and I called it transcendent, because it employed equations beyond all degrees; and I see that this method, almost alone of its kind, gives a method of determining whether particular problems of this kind are possible or not. Indeed I can easily prove in other ways, and also by the differential calculus more especially, the impossibility of general quadrature of the circle, or that no algebraical line can be given as its quadratrix. What I call algebraical lines are those that Descartes calls geometrical, and by quadratrices I mean all curves that, being described, will give the quadrature of any portion of a circle whatever. But the manner of finding the impossibility of any particular quadrature, for instance that of the whole circle, is known to me indeed in two ways, the one by the calculus of transcendent exponents, the other

[45][This I take to mean the principle that differentiation and integration are inverse operations; for it is practically certain that in November, 1675, he could not differentiate a product; otherwise, as previously argued, he would have verified his solution of the unfortunate equation, $x + y^2/2d = a^2/y$, which he gives as

$$(y^2 + x^2)(a^2 - yx) = 2y^2 \overline{\text{Log } y},$$

by differentiation, as he did with a previous solution that did not contain a product.]

[46][From this probably arose the first germ of the idea of the Quadratrix, in the sense used by Leibniz.]

[47][Substitute *Barrow* and Mercator in conjunction, and we have a feasible suggestion for explaining the first method of proof for the Arithmetical Quadrature of the Circle; the method that Leibniz does not seem ever to have divulged.]

by a certain new kind of calculus, embracing all cases, which has not entered the mind of any one before even in his dreams.[48]

"Here you have the story of some of my meditations...."

II.

From the correspondence between Leibniz and the Marquis de l'Hospital.

1694.

"I recognize that M. Barrow has advanced considerably, but I can assure you, Sir, that I have derived no assistance for my methods (*pour mes methodes*).[49] At the start I only knew the indivisibles of Cavalieri,[50] and the 'ductions' of Father Gregory St. Vincent, along with the "Synopsis of Geometry" of Father Fabri, and what could be derived from these authors and their like.[51] When M. Huygens lent me the "Letters of Dettonville" (or Pascal), I examined by chance[52] his demonstration of the measurement of the spherical surface, and in it I found an idea that the author had

[48][It is impossible for me to conjecture exactly which of his ideas is here referred to by Leibniz; for he calls a mere method by the name of "a calculus," and what we should call a dodge for some particular kind of example by the name of "a method." I think it may be possible that the "transmutation of figures" is referred to.]

[49][Notice that Leibniz says that he has not derived any help from Barrow for his *methods* (*je n'ay tiré aucun secours pour mes methodes*). This is less even than he might have said with perfect truth; for the *methods* of Barrow would have been a veritable hindrance to Leibniz's analytical development. Even when using the Differential Triangle method, and literals for the lengths of his lines, the whole of the working is geometrical in the examples of the method given by Barrow, and not analytical.]

[50][See Notes 35, 36.]

[51][Perhaps this is meant to include Barrow.]

[52][Notice the words "by chance" (*par hasard*); these seem to point to a conclusion that Leibniz read the Pascal in a very desultory manner; this conclusion gets corroborated by the extract given by Gerhardt under the heading III. It is worthy of remark that the "by chance," or "incidentally" (as I have rendered Leibniz's word *forte* in the letter to Tschirnhaus), is made to refer to Pascal. "*Forte Pascalius demonstrabat*," etc., i. e., "Incidentally Pascal was proving," etc. I think it may be asserted that Pascal missed absolutely nothing that was pertinent to *his purpose*; whereas Barrow certainly missed the opportunity of being the discoverer of the series for the inverse tangent, and thereby the quadrature of the circle, by not applying Mercator's method of division and integration to the result of one of his examples of the Differential Triangle method; as also after giving the method of "transmutation of figures" he missed those things to which it led.]

altogether missed; for I remarked that in general, by the same reasoning, the perpendiculars PC, when applied to the axis or set in the position BE, give a line FE, such that the area of the figure FABEF will furnish a development (*explanation*) of the surface formed by the rotation of AE about AB.

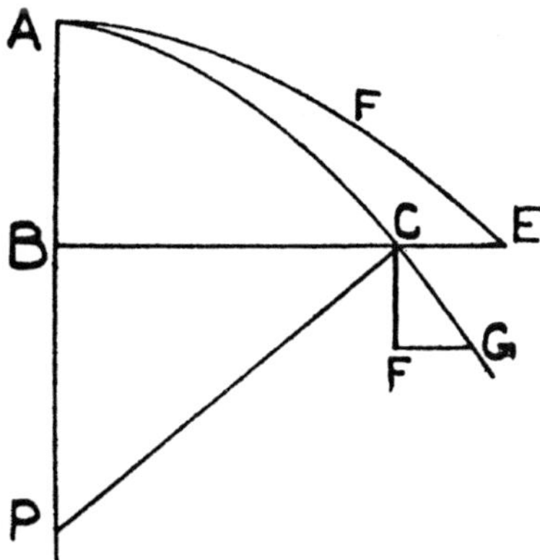

"Huygens was surprised when I told him of this theorem, and confessed to me that it was the very same as he had made use of for the surface of the parabolic conoid. Now, as that made me aware of the use of what I call the "characteristic triangle" CFG, formed from the elements of the coordinates and the curve, I thus found as it were in the twinkling of an eyelid nearly all the theorems that I afterward found in the works of Barrow and Gregory. Up to that time,[53] I was not sufficiently versed in the calculus of Descartes, and as yet did not make use of equations to express the nature of curved lines; but, on the advice of Huygens, I set to work at it, and I was far from sorry that I did so: for it gave me the means almost immediately of finding my differential calculus.[54] This

[53] [In a manuscript dated October, 1674 (see above, Chapter IV, p. 61), Leibniz is using x and y for the variable ordinate and abscissa; while in a manuscript dated August, 1673, he considers "the classification of curves laid down by Descartes." In this manuscript, according to Gerhardt, Leibniz has already constructed the "characteristic triangle," but Gerhardt does not give the particular variant that Leibniz uses in this manuscript. I believe that this will prove to be of the Barrow type, when reference can be made to the original; for the title of the manuscript is strongly suggestive of Barrow, being: *Methodus nova investigandi Tangentes....ex datis applicatis,* etc.; and Pascal's work does not mention tangents.]

[54] [That is, as the Characteristic Triangle, leading to integrations, is ascribed to the influence of the work of Pascal, so the Differential Calculus is ascribed to the influence of the work of Descartes. Is this the diplomatic characteristic in Leibniz peeping out? He is writing to a Frenchman, and attributes his work to the respective influences of two Frenchmen. Note that Leibniz goes on to state that the source of inspiration was summation of series by differences, suggesting the origin of the symbol dx.]

was as follows. I had for some time previously taken a pleasure in finding the sums of series of numbers, and for this I had made use of the well-known theorem, that, in a series decreasing to infinity, the first term is equal to the sum of all the differences. From this I had obtained what I call the "harmonic triangle," as opposed to the "arithmetical triangle" of Pascal; for M. Pascal had shown how one might obtain the sums of the figurate numbers, which arise when finding sums and sums of sums of the natural scale of arithmetical numbers. I on the other hand found that the fractions having figurate numbers for their denominators are the differences and the differences of the differences, etc., of the natural harmonic scale (that is, the fractions 1/1, 1/2, 1/3, 1/4, etc.), and that thus one could give the sums of the series of figurate fractions

$$\tfrac{1}{1} + \tfrac{1}{3} + \tfrac{1}{6} + \tfrac{1}{10} + \text{etc.,} \qquad \tfrac{1}{1} + \tfrac{1}{4} + \tfrac{1}{10} + \tfrac{1}{20} + \text{etc.}$$

Recognizing from this the great utility of differences and seeing that by the calculus of M. Descartes the ordinates of the curve could be expressed numerically, I saw that to find quadratures or the sums of the ordinates was the same thing as to find an ordinate (that of the quadratrix),[55] of which the difference is proportional to the given ordinate. I also recognized almost immediately that to find tangents is nothing else but to find differences (*differentier*), and that to find quadratures is nothing else but to find sums, provided that one supposes that the differences are incomparably small. I saw also that of necessity the differential magnitudes could be freed from (*se trouvent hors de*) the fraction and the root-symbol (*vinculum*), and that thus tangents could be found without getting into difficulties over (*se mettre en peine*) irrationals and fractions.[56] And there you have the story of the origin of my method...."

[At this point Gerhardt quotes his article, *Leibniz in London*, and a long passage from the *Historia*, in corroboration of the foregoing letters. I have omitted them as I have already, in my notes, pointed out the points of resemblance, and the slight differences, between the several accounts that Leibniz gives.]

[55] [In the manuscripts that we have had under consideration, Leibniz does not appear to have made any practical use of the Quadratrix.]

[56] [It is precisely this point which formed the really great improvement in the reckoning section of the infinitesimal calculus. It is just this improvement that is due to Leibniz in analysis, and to Barrow in geometry; although Leibniz did not accomplish anything of the kind until 1676 or 1677. Newton's method by means of series for fractions and roots does not bear comparison, let alone the futility of ascribing Leibniz's method to a perusal of Newton's work.]

III.

Extracts from the geometry of Dettonville or Pascal; with additions.
Ca. 1673.

1 2 3 4   If A, B, C, D, are quantities, their triangular
A B C D   sum, starting with A, is 1A, 2B, 3C, 4D.
   B C D   If BC is any straight line divided into any num-
      C D   ber of equal parts, and any weights, equal or unequal,
         D   are suspended at the points of division, and A is sup-
posed to be their point of equilibrium, it is necessary
that the triangular sum of the weights on the one arm AB should
be equal to the triangular sum of the weights on the other arm AC,
where the triangular sum on either side starts from the inner point
or from the side A. The reason is that the weights give an effect

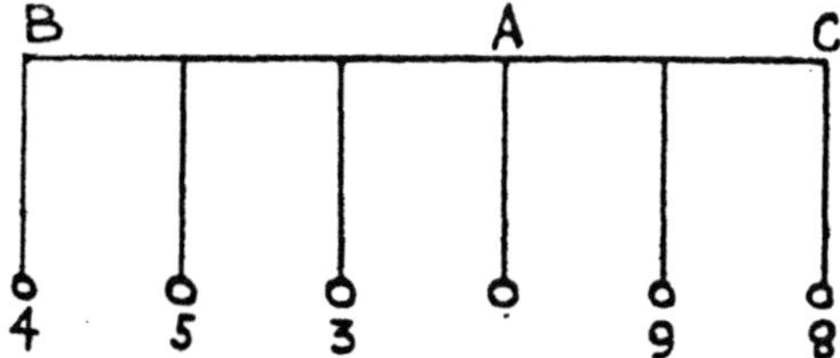

that is compounded of the ratio of the weights and their distances
from the center. But these distances, on account of the division
of the straight line or beam of the balance into equal parts increase
as 1, 2, 3, etc.

This is what Pascal says; to which I add the following remarks.

Even if the triangular sums on either side of the point are not
the same, that is if the two arms are not in equilibrium, yet the
moments will always be to one another as the triangular sums,
for the moments are always equal to triangular sums.[57] Hence the
far more general rule: If any straight line is divided into any number of equal parts, and weighted with any number of weights suspended at the points of division, and if any point of division is taken
to be A, then will the moments of the weights on the one arm BA
be to the moments of the weights on the other arm CA as the tri-

[57] [All that is any good in the following is to be found in Pascal; I think
this corroborates the suggestion I have made as to Leibniz's way when studying
a book. It looks here as if he had read about twenty pages of Pascal, and
about the same number of pages of Cavalieri's section on centers of gravity;
moved thereto probably or possibly by Pascal's remark "....the principle of
indivisibles, which cannot be rejected by any one having pretensions to rank
as a geometer." Then he proceeds to work out his own combination of the
two ideas, without bothering to see what else either of these authors had to
say on the matter.]

angular sums starting from that weight which is nearest to A on each side.[58] Also when any figure, i. e., a line, a surface, or a solid, can be put in such a position that a certain line in it can be taken as parallel to the horizon, that straight line can be taken as a balance, and all the points or all the straight lines or all the planes (where the points in the line are assumed to be placed horizontally, or lying in planes of these points set perpendicular to the horizon), may be considered as weights; and thus, if the quantity or progression of these weights is known, and consequently their triangular sum, then the center of gravity of the figure is known; not indeed its position in the figure, but its position in the straight line that has been taken. The center of equilibrium in the figure itself is of this nature: namely, that a straight line passing through it will cut the figure into two parts, such that on each side the triangular sums of the points, straight lines, or horizontals of the solids are equal to one another. Hence the center of gravity of the whole figure being found, the centers of gravity of arms of this kind supposable without the figure may be obtained; for, let the figure be A, and

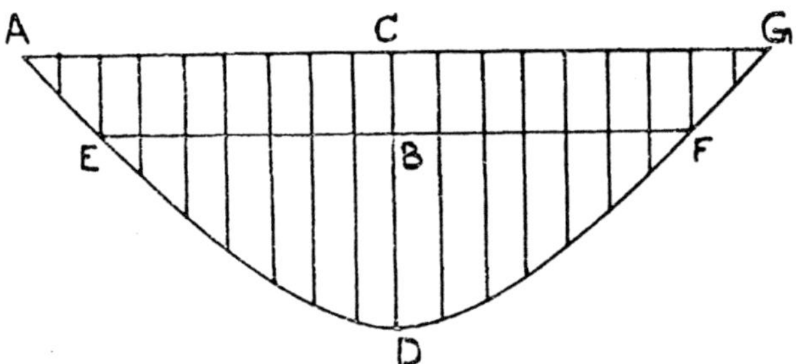

let there be taken a line parallel to the horizon in which is the center of gravity B, and suppose that the center of gravity of it is placed above a horizontal style or suspended by a thread: then it is plain that the figure will be in equilibrium. But if it is in equilibrium, then the straight line CD, drawn through the center of gravity, will cut the figure in such a fashion that the triangular sums on each side are equal; and if moreover another straight line perpendicular to CD is supposed to be divided into an infinite number of parts by the infinite parallels to CD, the triangular sums of the infinite rectangles on each side will be equal to one another, for by hypothesis the rectangles can be supposed to be suspended as weights from EF as a balance at the points of division (from which it is clear that

[58][Leibniz tacitly assumes that all the points are occupied; this is necessary for the success of the notion of triangular sums.]

the suspended weights need not necessarily be understood to be perpendicular to the horizon, but they may be parallel to it). This being the case, the position of the figure may be changed from the horizontal to the perpendicular, and AG become the balance; in which case it is clear that the point of equilibrium will fall at C, since the triangular sums are by hypothesis equal on each side of it. Hence, given the center of gravity of any figure, and assuming a balance either without or within the figure, to which the figure is supposed to be rigidly attached, the point of equilibrium can be found in it, by merely drawing a perpendicular to it through the center of gravity; for this will cut the balance in the point of equilibrium. On the other hand, if the points of equilibrium of two balances for the same figure are given, the center of gravity for the figure can be found (whether it is within or without the given figure; for sometimes the center will fall within the given figure; and sometimes without, as in the case of annular figures, or curved lines, or other incomplete things); that is to say, at the point of intersection of two perpendiculars drawn from those two balances toward the same parts, in the same plane, if the figure is a plane figure, i. e., if the balances are in one and the same plane; but if the two balances are not in the same plane, there is need for three. This is to be investigated.[59]

But the following is a better way: Suppose that the figure is first affixed to one balance, and let the plane through the common perpendicular be the balance and the horizontal be drawn through the point of equilibrium to cut the figure; then let the figure be affixed to another balance, and once more let another plane be drawn to cut the figure; the intersection of these two planes will give a straight line which will contain the center of equilibrium. If now a third balance is taken in addition, or a third plane, the point of intersection of all the planes, or the point in which the third plane cuts the line already found, will be the center of equilibrium. But if the figures are planes, then two balances and two perpendiculars are sufficient; and also if they are curved lines that lie all in the same plane.

Now it is worth while noting several things in those cases in which the balance is not divided into equal parts; for it may

[59] [Something very like this is indeed investigated fairly thoroughly in a manuscript dated October 25, 1675 (see above, Chapter IV, p. 65). Hence these extracts from Pascal were certainly made before that time, though probably not long before.]

happen that we may know in some way or other the sums of the weights and their progressions, but they are such that, when applied to the balance, they divide it into unequal parts; in that case the progression of the parts into which the balance is divided has to be investigated, as for instance if it is divided into parts that continually increase according to the squares or otherwise. Thus, if we wish to suppose that the weights are equal, while the balance is divided into parts that increase as 1, 2, 3, 4, etc., and yet that this case may come under the rule, we must proceed in this way. Suppose that that point of equilibrium is already found and that it is

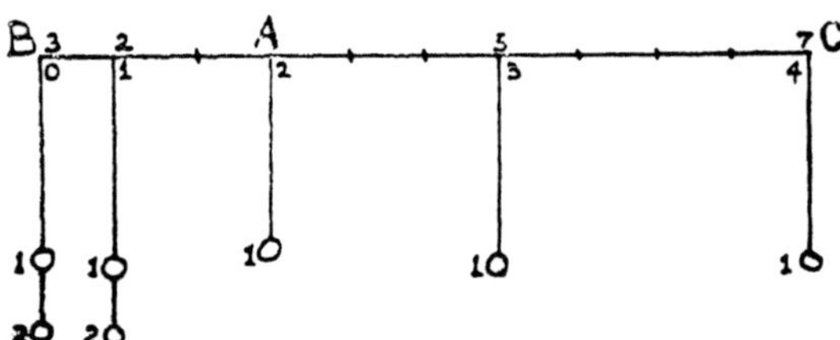

2, say; then it is clear that, starting from the point 2 assumed to be the center, the arms should be numbered, and that the point 1 should be marked with the number 2, and the point P with the number 3, and on the other side the point 3 should be marked with the number 3, and the point 4 with the number 7. Now, supposing that the weights are multiplied by the numbers of their own points or arms, it is necessary that the product obtained should be equal;[60] but if it is not, then another point must be sought (or something should be added to, or subtracted from, the weights; for instance, in this case, if the weights are 2, 3 should be supposed to be doubled, or in place of 1,1 we write 2,2 underneath, then there would be an equilibrium on each side, of 10). But to obviate the necessity of going through all the points, a formula should be sought; but if no known progression can be employed for the weights and the parts, a formula will be impossible; but when a known progression can be obtained, then a formula can be found as far as the nature of progression will allow. But the greater part of the difficulty will vanish in those cases in which the weights can be assumed to be equal. What is more, a very simple general rule has been found which is the reciprocal to that of Pascal, namely, that a point may

[60] [This is the rendering for *"productum fieri aequale"*; he probably means that what is produced on the one side, i. e., the sum of the moments on one side of A, should be equal to the sum of the moments on the other side. But this endeavor to obtain *something new* seems rather futile.]

be assumed such that the triangular sums of the numbers on each arm, always starting from the end and going toward the middle, are equal......[61]

[61][It would have been interesting to have seen what this simple rule was. Probably nothing more than the propositions given by Pascal as Prop. 1, 2, 3 of his method of the *balance*; this would corroborate my suggestion that Leibniz did not study Pascal very steadily or thoroughly (cf. Notes 37, 43, 52, 57 on pp. 216, 218, 220, 223 respectively).]

VIII.

CONCLUSIONS.

The notes and criticisms that I have made in these six chapters on the manuscripts of Leibniz may give the impression that I am an anti-Leibnizian. This is quite wrong. My prime object was to show, to the best of my power, that the charges of plagiarism brought against Leibniz by partisans of Newton, and indeed by Newton himself in the *Recensio* published in the *Philosophical Transactions,* were unfounded. I considered that the charges in the *Recensio* were perhaps the hardest to be answered, since they were not only direct charges, backed with circumstantial evidence, but they were also set forth very cleverly. Also I thought that the method of defense adopted by Gerhardt and other partisans of Leibniz did as much harm to him as the strongest attack of avowed opponents, such as Sloman. The weak case made out by Gerhardt is deplorable. Never surely did any man have such a glorious opportunity as Gerhardt, in the whole history of scientific controversies; surely there never was an advocate who left himself so open to the attacks of the opponents. Gerhardt starts with the theory that every single word of Leibniz represents gospel truth; and that it is almost blasphemy to doubt it; in consequence he is soon in difficulties when it comes to reconciling the varying statements of the sequences of events that are made by Leibniz at different times. But, once the idea is accepted that Leibniz, while perfectly reliable on the

general run of events, is unreliable when it comes to unimportant details, and then all difficulty disappears. I
therefore set out with the determination to break down,
if I could, the credibility of Leibniz as a witness in his own
defense, when it came to unimportant details; then to show
that he had opportunities for obtaining everything necessary to the development of the Calculus, that he could not
be expected to supply for himself by original work, without having need to know anything of the work of Newton;
then to show that these sources of information were set
out in a form far more suitable to the requirements of
Leibniz than the work of Newton; finally, to clinch the
matter, that the analogy of Leibniz's work was so close
to these sources, that it was idle to suppose that he made
use of any other sources. In other words, (i) the *Analysis
per aequationes* was unnecessary to Leibniz, (ii) Newton's
method of evading fractions and roots by means of infinite
series was clever, but futile for the needs of Leibniz when
developing an operational calculus.

The unreliability of Leibniz with regard to details may
be in some measure due to his apparently bad memory
(which is suggested by his habit of committing everything
to writing), and to passage of time. But in a far greater
degree it must be ascribed to the circumstances and characteristics of Leibniz. We know that he designed to com-
pile an encyclopedia of *all* science, and for this he considered not at all the nationality or the personality of the
discoverer or the author: all he was interested in were the
facts or principles discovered.

That he was unreliable with regard to details is proved
by the facts I have adduced:

i. the confusion between Mouton and Mercator in the
account of the assertion that he had been anticipated (see
above, Chapter III, p. 36, and Note 73, p. 37);

ii. the varied assortment of figures that he gives to illustrate how he found the Characteristic Triangle (see above, Chapter III, pp. 15 and 39, and compare them with the figures, given in the accounts quoted by Gerhardt in his essay "Leibniz and Pascal," on pp. 211, 217, and 221);

iii. the circumstantial detail of the context of the Archimedean measurement of the surface of the sphere being absent from the author he quotes;

iv. the several different accounts of the order in which he obtained his different books for study, and even the persons from whom he obtained them;

v. the error with regard to the time of the presentation of the copy of the *Horologium* (see above, Chapter III, p. 36, where, in the *Historia*, it is stated that he received it *before* he left for England on his first visit);

vi. the confusion as to the date at which he obtained his Barrow (see above, Chapter II, p. 20, where, in the Bernoulli postscript, he states that he found the greater part of his theorems anticipated in "Barrow, when his *Lectures* appeared");

and many other things, all unimportant details singly; but, when taken in combination, they show distinctly that we must only take Leibniz's word as accurately describing the *general* course of events.

Another characteristic of Leibniz seems to have been insistent at all times; he burned to distinguish himself as a discoverer of new things. I have suggested that there may have been an ulterior motive to this desire, namely, to get himself taken into the select circle of mathematicians who corresponded with one another. Thus, when he studied an author, and came across some new idea, he would break off his reading to follow that idea to the limit and exhaust all its possibilities, committing his results to writing, whether they were important or not; there is some evidence, too, that while doing this, he would refer to other authors who had discussed the point under consideration, before returning to his reading.

My motive in trying to show that he got everything from Barrow, *except his methods,* was to remove any charge of plagiarism; for, I consider that even if he had merely rewritten Barrow in terms of Descartes, adding his own notation for the sake of convenience, he would still have done a great thing, and would no more have been guilty of plagiarism from either Descartes or Barrow than Stephenson was from Watt, or Parsons from either of these. Leibniz's Calculus was his own, and would have been his own even on the supposition above. Lastly, it was not only more complete than that of Newton, in that it was an operational calculus, though it did perhaps miss the idea of rate; but also from an intellectual standpoint it was greater, in that it was developed, after its first principles were found out, as a practical theory, while Newton's was developed as a mere instrument for his own purposes.

Assuming, then, that Leibniz did not remember, or did not really care, what his text-books were, so long as he was not accused of using somebody else's *methods,* I will try and reconstruct the progress of his reading and his dis-coveries. His text-books were,

i. Lanzius and Clavius in algebra, and Leotaud for geometry, in his early youth; he also looked through, more or less without understanding them, Descartes and Cava-lieri's *Geometria Indivisibilibus.*

ii. On his return from London he brought back with him Barrow, some portions of which he had glanced at in London and on his journey; he obtained Pascal, St. Vincent, and Cavalieri's *Exercitationes Sex,* perhaps a little later than the others; besides these, Wallis and Mercator spe-cially.

He read portions of the Barrow afresh, and obtained the Characteristic Triangle, and found his general theorem from this; meanwhile he is also studying Descartes, and we have the materials for the manuscript of August, 1673. Probably he has had a look through Pascal during this time. He remembers the similarity between the complicated diagrams of Barrow and some of those of Pascal, and starts studying the *Traitté des Sinus,* in which he finds the second variant of the differential triangle that appears in the manuscript of October, 1674. Previous to this, however, his attention has been arrested by Barrow's proof of the inverse nature of the operations of finding a tangent and an area, and the analogy between this and sums and differences strikes him. He has also considered the examples on the differential triangle given by Barrow; one of them suggests the method of Mercator to him, he has already got an idea from Wallis of the summation of the several powers of the variable; he applies this to Barrow's expression, equivalent to

$$d(\tan^{-1}x)/dx = 1/(1 + x^2),$$

in modern notation, performs the division as Mercator had done, and obtains the series for the inverse-tangent by a summation according to Wallis, i. e., practically an integration. This answers the charge made by Newton that somehow or other he got this series from him or James Gregory. In the same way, he thought that he could obtain other series, but later found that it was beyond his power. We find in this manuscript of October, 1674, an attempt to get something out of an analogous series, the logarithmic series, showing that it is very probable that he has been studying Mercator during the interval between August, 1673, and October, 1674. And in the *Historia* he definitely states that he came upon the Arithmetical Tetragonism in 1674; so that I think that I have offered a reasonable suggestion as to the course his studies took so far. Also in the meanwhile he has been doing much work on series, and has invented his Harmonic Triangle. I now suppose that he completes his study of Pascal, is led by a remark in it to study the *Exercitationes Sex* of Cavalieri (he has already got some acquaintance with the *Geometria Indivisibilibus*, read as a youth), he does not find much in that to his liking, except the notion of moments. He breaks off his reading and proceeds to work out an application of Descartes's algebra to this new idea of moments, the result being the manuscripts of October and November, 1675; here he is led on to the introduction of the symbols for summation and differentiation, though as yet applied to series, and sums of powers. The consideration of the Quadratrix leads him to make a further study of Barrow; and he is led to x/d, by a consideration of Barrow's propositions on the inverse nature of the operations of integration and differentiation. This, combined with the analogy to the inverse nature of summations and differences, leads him to search for a reason why x/d should represent a difference such as he has considered to be denoted by dx. This at a later date necessitates the discussion of what the result of *operating* with d on a product or a quotient will be. Meanwhile the study of Barrow brings him to that proposition which gives the polar differential triangle; in it he perceives at once the method of "transmutation of figures." I now suppose that he appreciates Barrow more fully and begins to apply Cartesian geometry to Barrow's theorems; in a manuscript dated November, 1675, he attacked the problem of tangents, and in connection with it considered the method of Descartes. In the next manuscript that we have, dated June, 1676, he practically obtained the differentiation of the sine and the inverse sine; his figure, if he had given one, would have been the same as that of Barrow for the differentiation of the tangent. In July, 1676, he attacked the inverse-tangent problem, still considering the work of Descartes. I think, however, that his work on Barrow has taken effect, for from now on he includes the differential factor dx under the integral sign. This is the last manuscript before he went to London for the second time.

Thus, I take it that all Leibniz's work is the result of his own original methods on ideas that have been suggested chiefly by two books, those of Barrow and Descartes; at least, everything could have been suggested by these two books alone, except the notion of "moment," which came from Cavalieri. Thus it was unnecessary for him to have known anything about the work of Newton before he went to London for the second time. What he saw there may have had the effect of corroborating his own work; it could have had little other effect. The final polishing of his method I put down to a study of the Differential Triangle method of Barrow, which Leibniz perceived to be powerful, but found distasteful on account of the geometrical nature of the work.

BIBLIOGRAPHY.

ABBREVIATIONS

Cantor: Moritz Cantor, *Vorlesungen über Geschichte der Mathematik.* Vol. II, 2d ed., Leipsic, 1900; Vol. III, 2d ed., Leipsic, 1901.

De Morgan's *Newton*: Augustus De Morgan, *Essays on the Life and Work of Newton.* Edited with Notes and Appendices by Philip E. B. Jourdain. Chicago and London, 1914.

G., 1846: C. I. Gerhardt (Ed.), *Historia et Origo Calculi Differentialis a G. G. Leibnitio conscripta. Zur zweiten Säcularfeier des Leibnizischen Geburtstages aus den Handschriften der Königlichen Bibliothek zu Hannover.* Hanover, 1846.

G., 1848: C. I. Gerhardt, *Die Entdeckung der Differentialrechnung durch Leibniz, mit Benutzung der Leibnizischen Manuscripte auf der Königlichen Bibliothek zu Hannover.* Halle, 1848.

G., 1855: C. I. Gerhardt, *Die Geschichte der höheren Analysis. Erste Abtheilung* [the only one which appeared]: *Die Entdeckung der höheren Analysis.* Halle, 1855.

G. *math.*: C. I. Gerhardt (Ed.), *Leibnizens mathematische Schriften.* Berlin and Halle, 1849-1863.

Rosenberger: Ferdinand Rosenberger, *Isaac Newton und seine physikalischen Principien. Ein Hauptstück aus der Entwickelungsgeschichte der modernen Physik.* Leipsic, 1895.

Sloman: H. Sloman, *The Claim of Leibniz to the Invention of the Differential Calculus.* Cambridge, 1860 (English translation).

Barrow: *Lectiones Geometricae* of Isaac Barrow; translated, with notes, by J. M. Child. Chicago and London, 1916.

Zeuthen: *Geschichte der Mathematik im XVI. und XVII. Jahrhundert*; German edition. Leipsic.

Weissenborn: *Principien der höheren Analysis.* Halle, 1856.

Reiff: *Geschichte der unendlichen Reihen.* Tübingen.

Pascal: The edition of Pascal's works in 14 vols., in the series, *Les Grands Ecrivains de la France.* Hachette, Paris, 1914.

INDEX.

9 781603 860239